SYNTHETIC FUELS FROM COALS

SYNTHETIC FUELS FROM COAL

Overview and Assessment

LARRY L. ANDERSON

University of Utah
Salt Lake City

DAVID A. TILLMAN
University of Washington
Seattle

A WILEY-INTERSCIENCE PUBLICATION

JOHN WILEY & SONS
New York • Chichester • Brisbane • Toronto

Copyright © 1979 by John Wiley & Sons, Inc.

Library of Congress Cataloging in Publication Data:

Anderson, Larry LaVon.
  Synthetic fuels from coal.

  "A Wiley-Interscience publication."
  Includes index.
  1. Synthetic fuels. I. Tillman, David A., joint author. II. Title.
TP360.A52   662'.66   79-17786
ISBN 0-471-01784-1

Printed in the United States of America

10 9 8 7 6 5 4 3 2 1

For Tracy,
Camron,
Cyndy,
Nathan,
Belinda Sue, and
Darek

PREFACE

The six blind men and the elephant, perhaps, is the best fable to use when describing public perception of this nation's energy shortage. Some think it is merely a contrivance used to obtain higher prices. Others assume an apocalyptic stance—we'll soon run out of everything. Some would have the "greenhouse effect" heating up the earth and melting polar icecaps—as a result of burning fossil fuels and producing carbon dioxide. They seek a reduction in energy utilization. Others see the coming of an ice age. Some would regulate, stringently, all aspects of the energy business. Others seek complete laissez faire conditions. Through this maze of attitudes, opinions, and beliefs, corporate and government energy planners must wend their way. In this sea of confusion and contradiction, this nation and its economic sectors must chart a course of rational energy development.

Despite differing perceptions, "energy shortage" is now a ubiquitous term, and an obvious problem. The gap between domestic supplies and consumer requirements has been growing for a long time. This became obvious in 1973, when the oil embargo occurred, causing layoffs of over a million workers. Droughts in the Pacific Northwest that occurred in that year and in 1977 have resulted in the furloughing of thousands of workers in aluminum and related industries because electricity production was curtailed at the major hydroelectric generating stations. The natural gas shortfall in the winter of 1977 forced 2 million wage earners into unemployment for a short period, and half a million workers in joblessness for a longer span of time.

↑
con,

For some, the energy shortage has meant higher prices and inconvenience. For basic industry and its employees, shortfalls have carried with them lost production, profits, jobs, and wages.

The Federal government has commissioned numerous studies and created a host of agencies and organizations to deal with the problem. Right after the oil embargo, Project Independence Blueprint was developed. It emerged in 1974, alongside of the National Academy of Engineering's U.S. Energy Prospects: An Engineering Viewpoint. Contemporaneously the Energy Policy Project of the Ford Foundation published A Time to Choose. Since that time the Synthetic Fuels Commercialization Task Force has issued its four volume report for the U.S. Energy Research and Development Administration (now part of the Department of Energy). Other agencies commissioned and obtained similar massive tomes. Nuclear Power: Issues and Choices, for example, was the second major Ford Foundation energy policy report published.

While many studies were made, agencies entered the energy business in droves. But the 1977 natural gas shortage, and the now over $40 billion annual oil import bill have demonstrated that little has been accomplished. Perhaps the Federal government can do very little about energy. It is not organized for long-term planning. Its agencies tend to reflect specific needs rather than coordinated problem solving. Its efforts to operate in the marketplace have been less than successful.

As a result of such a situation, individual economic units must make their own energy policy at the plant level, the division level, and the corporate level. Industry and commerce need assured sources of energy supply to function. Their investors and employees depend upon management to make the right decisions. If manufacturers, warehousers, and other managers make wrong decisions, they will be penalized as no other segment of the economy is. They will go cold. In an age of relative energy scarcity, private enterprise bears the burden of the shortfalls. Industry shoulders this responsibility because it has more flexibility in raising capital and selecting alternatives than homeowners. It has fewer votes in the ballot box also.

Private enterprise establishes its energy policy in a manner very different from government. At the same time, however, both policy setting communities have two options to pursue—simultaneously. The first option is conservation. This approach increases the efficiency with which energy is used, and thereby reduces fuel consumption. Ideally it shifts the demand curve inward. This, however, is not a supply option. It adds not one Btu of new heat, not one erg of energy, to the system. The second option is finding adequate supplies of fuel—from coal, from wood, from whatever. Synthetic fuels are one key element in the supply option analysis, for they are capable of replacing the naturally occurring liquid and gaseous hydrocarbons now being used. The overutilization of petroleum and natural gas has created our increasing scarcity situation. Thus the synthetic fuels option merits critical analysis from both the user and producer points of view.

This text is structured to permit such an analysis. First, it develops the overall framework or context into which synthetic fuels must fit. It considers, in some detail, the reasons why alternatives to oil and gas are needed. It then surveys, briefly, what alternatives exist presently and for the next few decades. After establishing this context, the principles of coal conversion, plus the history of synthetic fuels production, are explored. The volume then focuses the analysis upon specific fuel types: low, medium, and high Btu gas; and heavy and light liquids. Synthetic fuels as petrochemical feedstocks also gain consideration. Finally, this text examines economic and environmental costs associated with synthetic fuel production.

The perspective or vantage point from which this analysis proceeds includes the broad areas of resource availability, reliability of domestic supply, and methods of production. Specific considerations include the nature of these fuels, their physical and chemical properties, their handling and combustion characteristics, their economic and environmental costs. These are fuel user oriented considerations, for potential fuel users ultimately will establish the micro and macro energy policies.

The result is an approach to the examination of synthetic fuel alternatives that provides managers and engineers with a method for assessing this series of options. It is designed for that purpose. It is presented as an aid to the development of energy policy in the private sector. For that reason, principles of fuel conversion and existing systems are stressed. Research systems and pilot plants are given less attention. Their potential role, particularly for the near and mid term, is not as well defined as the existing approaches.

Numerous individuals aided us in the development of this analysis, and their contributions should be gratefully acknowledged. These include Dr. B. D. Blaustein, at the Pittsburgh Energy Research Center; Dr. W. H. Wiser and Dr. A. G. Oblad of the University of Utah; J. Phinney, consultant to Consolidation Coal Co.; Dr. J. Boyd and Dr. E. T. Hayes of Materials Associates; and Dr. C. M. Mottley of the U.S. Department of Energy. These individuals and many others supplied information. They reviewed material as it was written. The entire project depended, in large measure, upon their contributions.

That many jobs have been lost because of inadequate supplies of energy in forms suited to present use capability is now known. That solutions can be found to this problem is now vitally important.

L. L. Anderson
D. A. Tillman

Salt Lake City, Utah
Seattle, Washington
July 1979

CONTENTS

SYNTHETIC FUELS
FROM COALS

CHAPTER 1

BEHIND THE TRANSITION TO ALTERNATIVE FUELS

1.1 FOR WANT OF A NAIL

Energy—the capacity to perform such work as powering automobiles and machines, generating electricity, heating homes and factories, cooking food, and accomplishing many other tasks—is so essential to human civilization that Herman Kahn, among others, considers learning to control fire the most significant advance of primitive man [1]. Energy, supplied in the form of liquid and gaseous fuels, has made possible modern agriculture. One farmer now can grow or raise enough food to feed 55 nonfarm persons. Those 55 persons can be engaged in the production of steel, automobiles, paper, television sets, and more. Those 55 people can run railroads, airlines, hotels, stores, restaurants, and other service establishments. Energy is the essential ingredient supporting all activity, directly and indirectly.

Substituting fuel energy—coal, oil, or gas—for human energy is the process of industrialization. It is the one process that permits manufacturing of large amounts of goods at prices most people can afford. Dr. E. Cook has calculated that in this decade farm labor, on average, costs $6,000/million Btu [2]. This contrasts with imported oil prices of $3.10 to $4.05/million btu, or coal prices of $0.75 to $1.50/million Btu. This substitution is based not only on price, but also on versatility. Human labor can be turned into mechanical energy, but it cannot be converted into the high grade thermal energy required for such processes as metals smelting and refining.

History documents the key role of energy; and the problems of fuel scarcity. In ancient Babylon, deposits of bitumen were discovered and quickly put to use providing fuel for the smelting

1

of metals and the heating of buildings. When the bitumen
deposits were exhausted, the Babylonian civilization disappeared
[3]. The rise of Crete as a producer and exporter of bronze
weaponry was based on the use of forest fuels [4]. That
society's decline resulted from deforestation. The salvation of
Great Britain's iron industry, when faced with similar deforestation, resulted from the successful substitution of coal-based
coke for wood-based charcoal. In Babylon and Crete, for want of
energy a civilization was lost. In England, a new source of
energy was found and used widely.

1.2 ENERGY IN THE UNITED STATES: THE CURRENT SITUATION

What of the United States? Presently this country consumes
close to 80 quadrillion Btu (quads)/year, importing increasing
amounts of oil to meet its needs. Thus it is useful to evaluate
ingrained patterns and reasons for energy demand, the historical
aspects of energy supply, and future demand-supply relationships
as they relate to consuming sectors and specific fuel sources.
From such a base of information, the energy alternatives available to industry can be explored.

1.2.1 Energy Demand in the United States

Energy demand, in the United States, is more dependent upon
the level of civilian employment than upon any other single
factor. From 1954 to 1962, workers were being added to the
civilian labor force at a rate of 637,000/year. From 1962 to
1975, 1.5 million workers were added to the civilian labor force
each year. The increase in net additions in 1962 stemmed from
postwar "baby boom" products entering the labor force. The more
rapid increase of women entering the labor force also began about
that time. During the entire period, energy consumption
increased by 1.57 quads for every million workers joining the
labor force. The relationship, depicted in Figure 1.1 is compel-
ling; it all but obliterates other statistical relationships [5].
From these data, relationships between energy consumption
and gross national product (GNP) can be defined. For the period
1947 to 1969, every billion dollar increase in GNP was accom-
panied by the expenditure of an additional 50,000 Btu. After
1969, the Breton Woods agreement and the subsequent dollar
devaluation changed the energy component of the relationship to
53,000 Btu [5]. Figure 1.2 defines this energy-GNP relationship
that has driven U.S. energy consumption to annual levels of near
80 quads.

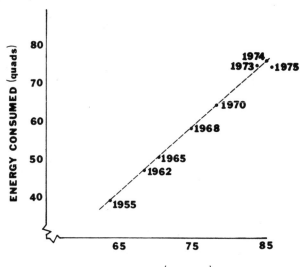

Figure 1.1. Energy and employment trends. As is shown
in this figure the creation of jobs requires the expenditure of energy. Every worker requires 1.55 billion Btu/
yr for all purposes. If employment is to continue to
grow, energy must be supplied to the economy in increasing amounts from a diversity of sources.

What lies behind, and drives, those aggregate relationships?
The Annual Survey of Manufacturers, 1974, reports that production
industries consumed approximately 39 quads in that year. The
largest consumers were iron and steel, petroleum refining,
organic chemical manufacturers. The 16 largest energy consuming
industries are listed in Table 1.1. They consume 56.7 percent of
the energy used in production. The remaining manufacturers,
however, consume almost 17 quads [6].

    Their energy needs can be expressed in Btu consumed/ton of
materials produced as well as aggregate totals or comparative
rank. For copper produced by the traditional mining-concentration-
smelting-refining method, the energy requirement is 112.3 million
Btu/ton of copper. For cementation copper, the requirement is
86 million Btu. Open hearth and basic oxygen steelmaking techniques require 26.1 and 27.2 million Btu/ton of product, respectively. Aluminum processing requires 245 million Btu/ton, while
magnesium production requires 358 million Btu/ton [2]. Taken as
a group, the primary metals industry consumes 8 percent of the

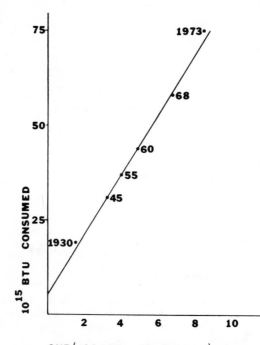

Figure 1.2   Energy consumed versus gross national product.
The long term relationship between energy and the gross
national product (GNP) as measured in constant dollars
is as strong as, and related to, the relationship between
energy and employment.

nation's energy supply. Of the industrial or nonmetallic minerals, portland cement stands out, consuming 7.6 million Btu/ton—
averaging almost 700 trillion Btu/yr for the total industry.
Plastics production requires from 45 to 135 million Btu/ton just
for process energy. With the continual rapid growth of plastics
in society, this fuel requirement is becoming highly significant
[2].

     Just bringing fuel from the deposits in the ground to the
point of consumption requires energy. For petroleum, 6 percent
of the energy extracted is required to get it to the refinery,
and another 11 percent—or 700,000 Btu/bbl—is required to refine
crude into its various products. In the case of natural gas,
7 percent of that extracted is needed to pump it to its final
locations. For coal, only 3 percent of the energy contained in

TABLE 1.1 Largest Energy Consuming Industries in
Manufacturing [6]

Rank	Industry
1	Blast furnaces and steel mills
2	Petroleum refining
3	Industrial organic chemicals, NEC.
4	Papermills (excluding building paper)
5	Paperboard mills
6	Cement, hydraulic
7	Industrial inorganic chemicals, NEC.
8	Primary aluminum
9	Nitrogenous fertilizers
10	Plastics materials and resins
11	Alkalies and chlorine
12	Organic fibers, noncellulosic
13	Glass containers
14	Cyclic crudes and intermediates
15	Motor vehicle parts and accessories
16	Motor vehicles and car bodies
(X)	All other manufacturing
(X)	Total manufacturing

the fuel is needed to mine it, clean it, and move it to the
ultimate user [2]. If this country is to maintain its present
standard of living, it must be willing to pay a higher energy
price in bringing energy to users in the most appropriate form.

1.2.2 Energy Supply in the United States

The history of energy supply in this country appears as a
series of cascading curves representing shifts from one primary
fuel to another. Wood provided the bulk of U.S. energy supply

until coal replaced it as the leading energy source in 1870.
Petroleum later replaced coal in this position. Natural gas
became a major factor after World War II and, although it never
surpassed oil, it did capture the second position from coal in
the late 1950s. One interesting and vital phenomenon can be
observed in this process. The fuel being replaced (e.g., wood
or coal) does not necessarily reduce its absolute contribution
of energy to the economy. Rather, the new fuel (e.g., coal,
petroleum) assumes the burden of supplying new energy users [7].

The reasons for these shifts include both convenience and
cost. One measure of convenience is the concentration of energy
per unit of fuel product. On the average, wood contains 16 million Btu/ton, coal contains 26 million Btu/ton, and petroleum
contains 40 million Btu/ton. Higher energy concentrations effec-
tively reduce transportation and storage problems for a given
fuel. Cost provides the second cause component of this shift.
Between 1950 and 1970, the wellhead price of petroleum declined
by 10 percent [8]. Natural gas prices were held at artificially
low levels, creating declining costs in constant dollars.

Petroleum prices began rising after 1970. The fourfold
price hike imposed by oil exporters in 1973, followed by subsequent increases, dramatically reinforced the situation. The 1978
decision of the OPEC ministers to increase oil prices by 14.5
percent during the year 1979 has added to the problem. Crude
prices of $23.50/bbl landed in the United States now exist.
During the revolution in Iran and the deposition of the Shah,
spot crude oil import prices reached $40/bbl. These prices are
far in excess of the $3.00 to $4.00/bbl oil prices of only a
decade ago. Natural gas prices have followed oil prices. Further, natural gas will be decontrolled by 1985. Yet today
petroleum and natural gas still provide over three-fourths of
our energy supply.

Domestic sources supply only about 50 percent of our oil
requirements; imports supply the remainder. In this fashion,
imports are acting like the traditional "new source," except for
one critical factor: domestic resources of oil and gas are supplying fewer and fewer Btu to the economy. Petroleum production
peaked in 1972, and natural gas production peaked in 1973. Production of oil is declining at an annual rate of about 4 percent,
and natural gas production is declining by over 5 percent/yr.
Of the petroleum that is produced, nearly 45 percent has come
from water flooding and other secondary recovery techniques
applied to mature fields [9].

The consequence is that imports are rising rapidly, supplying a critical component of U.S. energy demand. They now exceed
8.5 million barrels per day—or 17 quads/yr. Their cost to the

economy, in balance of payments outflows, is now in excess of $55 billion annually.

Industry obtains its energy in a fashion vastly different from society as a whole. Almost half of the energy used for manufacturing—49 percent—comes from natural gas, the fuel now experiencing the sharpest decline in domestic production [6]. Table 1.2 presents the energy consumption of industry by fuel type.

Because industry consumes more natural gas than any other fuel, it has been subjected to uncertainties of supply already. Throughout this decade, temporary disruptions have occurred. They reached a crescendo of sorts in the winter of 1977, when high demand driven by severe cold combined with inadequate natural gas supplies to create severe shortages of natural gas. Temporarily, 2 million workers faced unemployment. Half a million workers were out for over six months. Production was lost in all basic industries. The insulation companies felt the strain, since there was no way for them to keep up with conservation stimulated orders.

TABLE 1.2. Energy Consumption in Manufacturing Industry by Fuel Type, 1976 [6]

Fuel	Quadrillion Btu Consumed	Percent of Total Fuel Consumption
Natural gas	6.04	49
Oil	1.99	16
Coal	1.66	13
Electricity	2.17	17
Other	0.50	5
Total	12.48	100

1.3 ENERGY IN THE UNITED STATES: PROJECTED FUTURE DEMAND/ SUPPLY RELATIONSHIPS

Against that background, projections of future energy demand and supply can be considered. Of particular importance is their impact upon industry—the supplier of goods, services, and jobs. It is important, in this consideration, to avoid self-flagellation and breast-beating argumentation. It is essential to evaluate what is needed, where shortfalls occur, and how they may be addressed.

1.3.1 Future Energy Consumption

Based upon the well defined historical trends, U.S. energy consumption can be projected from a foundation of population and labor force projections. Again the work of Dr. C. M. Mottley can be used as a basis for analysis.*

Forecasts for energy demanded, for the year 2000, range from a low of ~70 quads to a high of 164 quads [10, 11]. The target presented is ~107 quads of energy. Based upon these data, energy demand will experience a 1.7 percent annual growth between now and 1985, but that rate of increase will decline to 1.2 percent by the year 2000 [5].

Industry may require an increasing share of energy in the years ahead, and for several reasons. The processing of lower grade ores and raw materials requires increased amounts of energy per ton of product produced. Since the turn of this century, production in the basic industries has increased fifteenfold, while energy consumption has risen 600-fold. Meeting environmental standards also consumes more energy. The switch to unleaded gasoline, for example, necessitated the use of 500,000 barrels of oil per day of additional energy in gasoline refining [2]. The steel industry, the nation's largest energy consumer, faces similar problems. Two independent studies—one by Research Planning Associates and one by A. D. Little—place increased energy consumption for environmental protection at 10+ percent for the steel industry. Commerce Department estimates

---

*These results are at considerable variance with the projections contained in the U.S. Bureau of Mines report, United States Energy Through the Year 2000 (revised). More recent reports, including one by A. Weinberg at the Institute of Energy Analysis, Economic and Environmental Implications of a U.S. Nuclear Moratorium, tend to support the forecasts made by Dr. Mottley although specific reasoning varies.

indicate that electric utilities will have to increase energy consumption by 9 percent to meet environmental laws. These energy expenditures are needed to run the electrostatic precipitators and scrubbers now required.[12]. At the same time, more people will be born, begin work, and seek the life style that is our culture. Industry will have to expand to meet this demand and, in the process, consume more energy.

Residential consumption will also increase on an absolute basis, but declining fertility rates plus a shift to more energy efficient multiple housing unit structures may reduce the share of fuel required by this segment of the economy. Forced energy efficiency in the automotive industry, within 12 years, will also make an impact.

1.3.2 Future Oil and Gas Supply

In the U.S. economy, consumption of oil and gas has become institutionalized. The transportation infrastructure is geared to these fuels. Boiler designs for small and medium sized applications are based on their properties. Thus an analysis of future U.S. energy supply must begin with oil and gas.

The Committee on Nuclear and Alternative Energy Systems of the National Academy of Sciences (CONAES), addressing future production of oil and gas, foresaw nothing but irreversible decline—even under wartime footing conditions. Under an extension of present conditions or moderate enhancement induced by a concrete government policy with modest incentives, the declines will be severe. Their forecast through the year 2010 demonstrate that oil can only slow down the rate of declining production. Studies by Herman et al. [13] and by the Congressional Research Service, Library of Congress [14, 15] have reached the same stark conclusions.

Exacerbating this decline has been the partial and/or total protective withdrawal of several millions of acres of Federally owned lands. Government acts have reserved those lands either for single purpose development (i.e., oil shale development) or as wilderness or recreational land. These withdrawals put added burden on other prospects and on existing fields. Disruptions in the leasing of the Outer Continental Shelf and other offshore areas has created similar problems for oil and gas production. Their unavailability increases the rate of decline at which available oil and gas from domestic deposits is produced.

1.3.3 The Consequence for Industry

The consequence of this pattern of decline will be increasingly intensive competition for the remaining supplies of oil and gas. The competition, to date, has been political. It will remain so. When one considers that 39.5 million homes are heated with natural gas and 14.8 million homes are heated with oil, it is easy to see why industry cannot win the political fight [16].

The consequences of this competition are already appearing. The American Gas Association forecasts that, by 1990, natural gas available to industry will be cut 30 percent. They predict that the western north central states of Minnesota, Iowa, Missouri, Kansas, Nebraska, and the Dakotas will lose over 40 percent of their industrial gas [17]. The Federal Energy Administration estimates are more bleak. They predict that, by 1985, no natural gas will be available for industries located in the midwest [18]. The decline of oil and gas production, and the consequences of that decline for industry, are real—and brutal.

Industry will have to find alternatives. Already it is losing production to natural gas curtailments and at increasing rates. Energy shortage layoffs are not uncommon. The Texas Railroad Commission has said, as regulatory policy, that natural gas will be out from under utility and industrial boilers by 1985. The Federal Energy Administration has ordered 79 electricity generating stations and over 100 industries to convert from oil and gas to coal. Petroleum and natural gas as industrial fuels will be scarce.

Petrochemical companies face the most serious dilemma. Those firms offer the strongest industrial competition for remaining petroleum supplies. Their access to petroleum is particularly important, for oil and gas are materials, as well as energy, to the petrochemical industry. Plastics have become indispensable to the U.S. consumer economy. The industry's annual growth rate, to the year 2000, is forecast at 8.6 percent. At that time the plastics industry will contribute 7.2 percent to the GNP. At that time this country will be consuming 1.5 million tons of plastics to meet a variety of essential and desirable purposes [19]. With massive domestic investments, these firms may follow the route of Union Carbide Corporation. That firm is developing a process to convert crude oil directly into feedstocks—permitting the use of higher priced liquid hydrocarbons as the price for maintaining domestic operations [20].

Industry is already facing the problems of finding substitute fuels—bearing the burden of shortfalls. Not only has it lost the political battle over supplies, but it also has more ability to raise the capital necessary for conversion than individual homeowners.

Industry does have several options available, as Chapter 2 explores. In addition, the manufacturing, utility, and commercial communities can continue their conservation programs. It must be remembered, however, that conservation is not a supply option. It reduces the energy required per unit of output or per job created. It does not supply one additional Btu. Thus after having taken a cursory view of the cause of our energy problem, this book takes a brief look at the basic alternatives available to industry. It examines, in detail, the coal conversion alternatives of low, medium, and high Btu gases; and heavy and light coal liquids.

REFERENCES

1 Herman Kahn, William Brown, and Leon Martel, The Next 200 Years, New York: William Morrow and Company, Inc. (for The Hudson Institute), 1976.

2 Earl T. Hayes, "Energy Implications of Materials Processing," Science, Vol. 191, No. 4228, Feb. 20, 1976, pp. 661-665.

3 David Cass-Beggs, "Energy and Civilization," in Proceedings-International Biomass Energy Conference, Winnipeg, Manitoba: The Biomass Energy Institute, 1973.

4 H. G. Cordero and L. H. Tarring, Babylon to Birmingham, London, England: Quinn Press, Ltd., 1960.

5 Charles M. Mottley, "How Much Energy Do We Really Need," in Fuels and Energy From Renewable Resources (Symposium Volume), David A. Tillman, Kyosti V. Sarkanen, and Larry L. Anderson (Eds.), New York: Academic Press, Inc., 1977.

6 Annual Survey of Manufacturers 1976: Fuels and Electric Energy Consumed, Washington, D.C.: Bureau of the Census, May 1978.

7 H. C. Hottel and J. B. Howard, New Energy Technology—Some Facts and Assessments, Cambridge, Mass.: The MIT Press, 1971.

8 U.S. Bureau of Mines, Commodity Data Summaries 1976, Washington, D.C.: U.S. Department of Interior, 1976.

9 Personal discussions with Dr. Charles Mankin, Director, Oklahoma Geological Survey, Norman, Oklahoma.

10 Interim Report of the National Research Council Committee on Nuclear and Alternative Energy Systems, Washington, D.C.: The National Academy of Sciences, January 1977.

11 Walter G. Dupree, Jr., and John S. Corsetino, <u>United States Energy Through the Year 2000</u> (Revised), Washington, D.C.: Bureau of Mines, U.S. Department of the Interior, December 1975.

12 "Does Pollution Control Waste Too Much Energy," <u>Business Week</u>, March 29, 1972, p. 72.

13 Stewart W. Herman, J. S. Cannon, A. J. Malefatto, <u>Energy Futures</u>, L. H. Orr (Ed.), Cambridge Mass.: Ballinger Publishing Co., Chap. 12, pp. 474-476, 1977.

14 Congressional Research Service, Library of Congress, <u>Project Independence: U.S. and World Energy Outlook through 1990</u>. Washington, D.C.: U.S. Government Printing Office, November 1977.

15 Herman T. Franssen, <u>U.S. Energy Demand and Supply, 1976-1985; Limited Options, Unlimited Constraints</u>. Washington, D.C.: U.S. Government Printing Office, November 1977.

16 Battelle Columbus Laboratory, <u>Evaluation of National Boiler Inventory</u>, Washington, D.C.: Environmental Protection Agency, October 1975.

17 Andy McCue, "Predict Gas Cut to Industry by '90," <u>Energy User News</u>, October 18, 1976.

18 Paul Schaffer, "FEA On Midwest: No Industry Gas by '85," <u>Energy User News</u>, March 21, 1977.

19 R. L. Glauz, Jr., et al., <u>The Plastics Industry in the Year 2000</u>, Palo Alto, Calif.: Stanford Research Institute, April 1973.

20 R. S. Wishart, "Industrial Energy in Transition: A Petro-chemical Perspective," <u>Science</u>, <u>199</u> (4329), February 10, 1978, pp. 614-619.

CHAPTER 2

THE RANGE
OF ALTERNATIVES

2.1 INTRODUCTION

Industry must make the transition from oil and gas, demonstrated in Chapter 1. It has relatively more flexibility than the residential sector. The fuels that industry can turn to and the role of synthetic fuels among those fuel options, therefore, become critical questions. Synthetic fuels are a series of alternative energy sources, but they exist in a much larger range of fuels available or potentially available to industry. This chapter does not pose as a detailed examination of all nonsynthetic options. Rather it offers a survey investigation designed to establish a reasonable context within which synthetic fuels may be analyzed. It is particularly important to establish this context, for synthetic fuel energy alternatives do not exist in a vacuum. It is important, also, because this survey develops certain observations that may be useful in broader analyses.

In order to assess the issues of fuel transition, the energy utilization patterns within industry must first be analyzed. Then near- and mid-term options can be evaluated and related back, ultimately, to those established patterns. The overall guiding principle underlying this transition is best expressed in a letter from Dr. J. Boyd to Mr. W. K. Davis, Vice President of the Bechtel Companies:

"The facts clearly indicate that there has to be a major change in the patterns of usage of energy in order to maintain a healthy economic society. This is because the present delineated and known reserves of the basic energy raw materials exist in a far

13

different ratio to each other than the current rate of usage. The issues we face then are those of taking what energy sources are available to use and creating a new supply system based more nearly on the ratio of availability [1]."

2.1.1   Industry's Fuel Utilization Constraints

Since World War II, the manufacturing and some utility sectors of the economy have institutionalized the use of natural gas and oil in their boilers and kilns. In the initial conversion to these fuels, the coal handling equipment, including ash collection and removal systems, was abandoned. Then facility expansions occurred where the coal pile had been.

The institutionalization of oil and gas was caused by lower capital and operating costs associated with the use of these fuels. It proceeded unabated through the 1960s and early 1970s. It was aided not only by artificially low gas prices, but also by environmental regulations. It was hardly affected by the oil embargo or other shortage warnings. Figure 2.1 depicts oil and gas industrial boiler capacity sales trends since 1961, as calculated by the American Boiler Manufacturers Association [2]. Figure 2.2 depicts oil and gas fired boiler installations in commercial and light manufacturing facilities [2]. Taken together they illustrate an institutional constraint impeding the conversion. Since 1960, several hundred billion dollars have been invested in liquid and gaseous fuel boilers—with lesser, but significant amounts invested in kilns operating on the same fuels.

Today the industrial and industrial-commercial boiler population exceeds 40,000 units (not including utility boilers). These industrial and commercial-industrial boilers purchased by manufacturers have a long life. For tax purposes, the Internal Revenue Service permits depreciation over 25 years. For practical purposes, the ABMA estimates a life of about 30 years. Some can last as long as 40 years. Thus, the boilers bought between 1960 and 1965 will be operating until 1985 to 1995. In making the transition away from oil and gas, some alternatives would force premature abandonment of these boilers, and the capital investment that they represent. In a society where increasing competition for capital exists, energy alternatives that preserve the usefulness of existing boilers take an added significance.

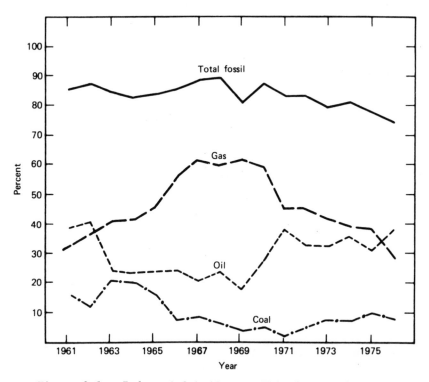

Figure 2.1. Industrial boilers. This figure shows
the percent of industrial boiler capacity sold by year
as a function of fossil fuel type. The 25,000+ lb/hr
boilers shown are designed primarily to use oil and
natural gas. Thus there is a strong potential demand
for replacement synthetic fuels. (Source: American
Boiler Manufacturers' Assn.)

15

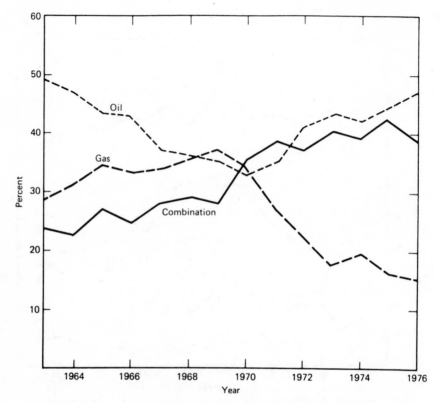

Figure 2.2. Commercial-industrial boilers. Annual
sales of the smaller package boilers accentuates the
need for oil and natural gas substitutes. Virtually
all of the smaller sized boilers are designed to fire
clean, convenient fuels. (Source: American Boiler
Manufacturers' Assn.)

2.1.2   Approach to Alternative Fuel Analysis

In order to evaluate alternatives available to industry, energy forms are considered in this sequence: other combustible fuels—imported oil, coal, oil shale, and organic wastes; noncombustible mineral fuels—uranium and geothermal energy; the "eternal fuel"—solar energy; and the converted fuels—electricity and coal based synthetics.* It can be observed, generally, that all of these fuels will play some role. In defining, briefly, the role of these key alternatives, the place of coal-based synthetics can be established.

2.2   THE ANALYSIS OF ALTERNATIVE FUELS

All fuels are considered here on a relatively uniform basis. They are analyzed in terms of their short- and long-term availability, their utility in meeting space and process heating requirements, their costs of use, special technical problems, and the ability (or inability) of each option to interface with existing capital investments such as the present boiler population.

2.2.1   Combustible Fuel Alternatives

The basic combustible fuel alternatives for industry include imported oil, solid coal, oil shale, and a variety of cellulosic materials that are residues of timber extraction, product manufacturing, and materials utilization.

1.   Imported Oil.   Some grades of oil may be available from foreign and domestic sources, to a limited extent, to industry. However, depending upon foreign oil is somewhat like depending upon domestic oil. The resource base is limited, the expected life span of those resources is relatively short, and the competition for those resources is becoming fierce. Already this country by 1978 was importing about 8.5 million barrels of oil per day—over 17 quadrillion (quads) Btu per year. It is

---

*The principle of classifying synthetic fuels with electricity, rather than oil and gas, is best presented by Dr. E. Reichl in a paper given at the Boston Regional Meeting of the American Association for the Advancement of Science, Feb. 20, 1976, "The Status of Coal Conversion."

generally recognized that this level of imports is unhealthy for
the U.S. economy.

More critical for individual industrial and utility planning
purposes is the general recognition that world oil could become
critically short after the mid 1980s. This will be caused by production capacity limitations of the oil producing nations. Oil
producing nations would have to increase their production capacity
from 29 MMBD to 50 MMBD to satisfy present growth trends. Such
capacity expansions could be physically impossible. Competition
from other countries with fewer energy options than the United
States also limits our access to world oil [3]. The potential for
supply disruptions is increasing, and more disruptions carry the
implication of lost production, profits, jobs, and wages. Even the
optimistic Ford Foundation Study [4] offers admonitions concerning
importing more oil. Thus, at best, this option is of extremely
limited utility.

2. Coal Combustion. Coal combustion is a viable alternative to
oil and gas, if one is willing to pay the cost for converting to
this solid fuel. Currently, recoverable reserves of coal in the
United States exceed 280 billion tons, or 6,000 quads and this
only includes reserves recoverable by present technology, at
depths of less than 1000 ft and at thicknesses of 28 in. or
greater. Table 2.1 presents these recoverable reserves by coal
type. Further, only one region in the country lacks coal reserves,
New England. All other parts of the nation have indigenous coal
supplies as Table 2.2 demonstrates.

Coal production has been increasing since the mid 1960s, and
present mines have the capacity to produce 15 percent more than
they are now supplying to the economy [5]. Production could rise,
almost immediately, to over 780 million tons and, by 1985, reach
1.0 to 1.2 billion tons [4]. Presently coal production is constrained to less than 700 million tons by the lack of customers
willing to purchase and burn it. Coal production is demand constrained.

Many firms already have switched from oil and gas to coal
combustion systems. The cement industry, for example, has invested over $125 million since 1970 to make the conversion. Several
brick companies and paper mills have turned to coal. Taconite
(iron ore) producers, with the help of the U.S. Bureau of Mines,
are beginning to turn from imported natural gas (of Canadian ori-
gin) to coal [6]. These companies and others have created a
growing industrial (nonutility, nonmetallurgical) demand for coal,
that now exceeds 80 million tons/yr. In just one year, January
1976 to January 1977, this industrial coal demand increased by 19
percent according to U.S. Bureau of Mines Data. Firms converting
to coal are willing, and able, to pay the price for fuel security.

TABLE 2.1 Recoverable Reserves of U.S. Coal by Type of Mining and Type of Coal [5][a,b]

	Type of Mining					
	Underground		Surface		Total	
Type of Coal	MM Tons	1×10^{15} Btu	MM Tons	1×10^{15} Btu	MM Tons	1×10^{15} Btu
Anthracite	3,650	94.8	80	2.0	3,730	96.8
Bituminous	96,200	2,498.6	34,500	895.6	130,700	3,394.2
Subbituminous	50,100	1,002.1	57,860	1,157.2	107,960	2,159.3
Lignite		N/A	23,940	383.0	23,940	383.0
Total	149,950	3,595.5	130,700	2,437.8	266,330	6,033.3

[a]Reserves based on deposits less than 1000 feet below the surface and seams thicker than 28 inches.

[b]Data from other sources may be slightly different. For example, U.S.G.S Bulletin 1412 (Jan. 1974) and U.S.B.M. Information Circular 8531 show comparable reserves at 148.02 billion tons for underground and surface reserves at 84.16 billion tons for a total of 232.18 billion tons. The National Petroleum Council gives "recoverable reserves" minable by surface mining methods as only 45.0 billion tons (Coal Availability, U.S. Energy Outlook, Report prepared for U.S. Department of Interior).

TABLE 2.2 Distribution of Recoverable Reserves by Census Region and Type of Coal [5]^a
(expressed in 10¹⁵ Btu)

Region	Type of Coal				
	Anthracite	Bituminous	Sub-bituminous	Lignite	Total
New England	0	0	0	0	0
Middle Atlantic	93.3	320.1	0	0	413.4
East North Central	0	1,425.6	0	0	1,425.6
West North Central	0	222.4	0	223.5	445.9
South Atlantic	1.8	673.0	0	0	674.8
East South Central	0	441.1	0	14.0	455.1
West South Central	1.2	29.9	0	44.9	76.0
Mountain	0.4	296.6	1,952.0	96.5	2,345.5
Pacific	0	29.8	163.1	4.1	197.0
Total	96.7	3,438.5	2,115.1	383.0	6,033.3

^aSee Table 2.1 for limitations of depth and seam thickness.

The cost of exercising this option is not inconsiderable. In 1976 dollars, the capital cost for the capacity to raise 1 lb steam/hr is $40. When pollution controls are added, this capital cost is $60/lb steam/hr [7]. In contrast, the capital cost of a liquid or gaseous fuel boiler is $15/lb steam/hr [7]. Further, while it may take a year to install a boiler fired by liquid or gaseous fuels, it takes three years to erect a coal fired unit. Finally, boiler fabrication and construction capacity does not exist to handle a massive influx of new boiler orders [7]. Thus, while converting to coal is a very real option, it is a most expensive and time consuming one. It involves not only the costs previously described, but also the penalty of forced, premature retirement of the existing boiler capacity. This penalty may be unacceptable.

3. Oil Shale. The vast resources of oil shale make this option potentially of great importance. The Green River formation in Colorado and Utah contains 418 billion barrels in identified resources of 25 to 100 gal/ton [8]. This amount of material is equivalent to over 2400 quads of energy. The total energy contained in identified and hypothetical resources of the Green River formation is over 14,000 quads [8]. The problems associated with developing this resource are formidable, however. In addition to the well-known water issue, developers have to construct entire towns to house workers, develop systems for handling solid residues of processing shale, and cope with stringent air quality regulations.

The magnitude of the resource base, however, stimulates continued efforts to develop an economically sound recovery system. Recent successes in developing modified in situ processes, plus a slight relaxing of environmental standards, has again created optimism that such a recovery system can be achieved [9]. Assuming the achievement of an economically and environmentally acceptable process the most optimistic forecasts call for the production of about four quads of energy by the year 2000 [9]. It will be most sought after by residences and petrochemical industries as well as industry. Thus oil shale development as an industry and as a source of energy alleviates some problems, but does not affect the need for coal conversion.

4. Lignocellulosic Materials. Wood and wood waste, agricultural residues, industrial waste, and municipal solid waste combine to form the final category of combustible fuels to be considered by industry. Many firms already are using these fuels, including General Motors, Xerox, John Deere, General Electric, Goodyear Tire, Eastman Kodak, and the entire pulp and paper, lumber and plywood (silvicultural) industry. These residues are now supplying the

industrial community with some 1.8 quads of energy annually
[10].

What remains annually as collected and available but unused
residue, however, approach 150 million tons—or about 1.5 quads
per year. The total generated annually, in concentrated and dispersed forms, is about 1.1 billion tons—or 11 quads [11]. Most
of this dispersed material could be used only at prohibitive
costs.

Because the potential of this energy alternative is relatively small does not mean that it should be overlooked. Technology exists to use this energy resource. More and more companies are employing this option, but the restriction on fuel
availability implies that few firms outside the silvicultural and
cane sugar industries can rely upon residues to supply more than
a modest fraction of their energy requirements.

In addition to supply limitations, this option poses some
investment difficulties. Unless waste is converted into gaseous
forms, new boilers are required at a cost of premature obsoles-
cence for existing equipment. Boilers to burn these fuels in
solid form have a capital cost of $33 to $40/lb of steam/hr of
capacity. Further, their efficiency in generating steam is in
the vicinity of 70 percent [12]. This is significantly lower than
the 80 to 85 percent efficiency of coal, oil, and gas-fired boilers. For those firms purchasing gaseous waste based fuels, the
price is nearly equivalent to the cost of distillate oil.

For many companies, however, this alternative is very useful
as a means for waste disposal and partial or total energy supply.
In most cases, the synergy between environmental protection and
energy supply makes it economically attractive.

2.2.2   Noncombustible Mineral Energy Sources

Two noncombustible mineral energy resource options exist:
uranium or nuclear power and geothermal steam and brines. Both
nuclear and geothermal energy have been utilized for electricity
generation. Because both can raise steam, they have potential
application in manufacturing, despite the fact that neither can
interface with existing fuel utilization equipment. Thus their
potential availability, at least, must be explored.

1. **Nuclear Power.** Nuclear power, fueled by enriched uranium,
relies upon increasing discoveries of that metal. Today, reserves of uranium are about 640,000 tons, with an estimated additional 60,000 tons of byproduct production from phosphates and
copper also available. Thus, this country is reasonably certain
that 700,000 tons of uranium can be made available, in the near
future, for economic growth purposes. For the most part, that

uranium will be needed to fuel the 127 gigawatts (GWe) of capacity already built, in design or in construction. The margin between those requirements and reserves is a scant 1.5 percent [3].

Growth beyond 127 GWe depends upon finding resources and converting them into reserves and ultimately into producing properties. There is considerable debate on whether this is practical in this century. The Ford Foundation study [3] is optimistic concerning future growth of the present lightwater reactor industry. Arguing that geologists, in times past, have been too conservative in estimating resources and reserves, that policy planing document estimates that the resource base of 3.7 million tons postulated by ERDA is conservative. Their arguments in support of this contention are both historic and economic. This position is not universally accepted. The National Academy of Sciences' CONAES panel on uranium, arguing from geological evidence and reasoning, stated it is quite probable that only 1.1 million tons of resources exist beyond the identified and inferred reserves [13]. Such a level of resources could only support an additional 180 GWe of capacity. The CONAES panel went further, arguing that the time required to find, prove, and develop these resources would seriously limit uranium availability. Further, they foresaw constraints on production capacity.

There is a second problem associated with nuclear power, that of construction lead times caused by the complexities of design and the intricacies of the regulatory process. It now takes over 12 years to get a nuclear power station on line. Such a construction period adds to the capital cost, increases the investment risk, and discourages would be nuclear users. Thus plants committed today will not be available until 1991—or later.

Despite such limitations, the use of uranium is growing—as it has to in order to fill some of the void created by declining domestic supplies of oil and gas. Between now and 1985, 125 new nuclear power plants are scheduled to come on line. Utilities building such plants are taking an expensive option; for the capital requirement to build a nuclear power plant is 130 percent of the capital requirement for a coal plant. Further, the minimum economic size is now 1100 megawatts (MWe) of capacity. This option, then, is available only for massive energy consuming facilities.

Thus industry cannot prudently plan on additional expansion of the nuclear industry above present plans to fuel both electrical and manufacturing complexes. Nuclear power has been proposed for steelmaking. The emergence of the pioneering marriage between Consumers Power and Dow Chemical in Midland, Michigan—to use nuclear power to produce both electricity and process steam—

exists. Neither case, however, mitigates the serious resource
and production constraints inhibiting future growth of this energy
form.

2. Geothermal Energy. Geothermal energy, broadly speaking, is
heat migrating outward from the center of the earth. For practical energy planning purposes, however, it consists of superheated
steam and liquids trapped in geological structures that when
tapped, yield steam and/or liquids for harnessing; or hot dry
rock. Identified reserves capable of generating electricity offer
only some 79.4 quads [14]. While it is recognized that geothermal energy resources, for all practical purposes, are nearly infin-
ite, the 79.4 quads (equal to 13.7 billion barrels of oil) is all
that can be planned on.

     Two problems are paramount in exercising this option—reduc-
ing its utility to that of a supplementary fuel of modest propor-
tions. It is regionally concentrated in the southwest—predominantly in California, Idaho, and Utah. Thus it is only available
to utilities and industries in a small geographical section of
the United States. A second, more serious, limitation is the
problem of extracting the fluids and using them. Dissolved salts
and metals are in the geothermal brines. When the brine is
brought out of its environment it corrodes the well pipes. Further, it builds a scale on the pipes that, ultimately, constrict
the flow of liquids from the deposits to the power plants [15].

     Geothermal development is proceeding, but it is unlikely to
ever become a primary energy source. For those who can take
advantage of it, it should not be overlooked. For others, energy
must come from other sources if their role in the economy is to
continue.

2.2.3 Eternal Energy—Solar Energy

     Solar energy is virtually eternal—as long as the sun shines,
solar energy will be available. As one observer put it: "It's
free! The only cost is that of collection." Therein lies the
problem, for solar energy is dilute. One can consider wind,
tidal, or ocean thermal energy as forms of solar energy; but they
are in the future. The technology available for present economic
planning purposes is the flat plate collector, although refine-
ments and variations are being developed.

     Already, several hundred homes plus several schools, commer-
cial establishments, and factories are heated with the sun's rays
[16]. Companies using solar heat include General Electric, John
Deere, and The Gump Glass Company of Denver, Colorado. The Oak-
mead Industrial Park of Santa Clara County, California, is
designed to use this energy resource [16]. A solar collector

industry now exists that has passed the production of 700,000 sq ft of collectors per year—and has passed the $10 million sales mark [17]. Although the solar industry is still small, it is growing.

Solar energy can supply domestic heat and hot water, but the collectors normally supply water to the heating system at 140 - 180°F (60 - 82°C). This is far below the temperature requirements of process heat. Industrial process heat solar systems have been built, experimentally. The first solar furnace for manufacturing was built in Egypt, in 1904. Such systems, however, can only be considered long range when their capital cost and reliability are compared to more traditional energy systems. Further, when the sun doesn't shine, a back-up system must exist if industry and commerce are to continue. Thus this energy source is again a supplementary alternative that merits examination but that has far from ubiquitous application. It is yet another partial, but limited alternative for firms seeking to decrease dependence upon and ultimately detach themselves from oil and gas.

2.2.4 Converted Energy—Electricity and Synthetic Fuels

In searching for an alternative to oil and gas, manufacturers have one final set of options: conversion fuels—electricity and synthetic fuels. While these are not primary energy alternatives, they are real options available to industry.

1. **Electricity.** Electricity (other than that which is self-generated) has been used increasingly by manufacturers in their search for ways to meet environmental regulations. It can supply space heat or process heat, and it can be used in metals reduction and fabrication. From both environmental and capital cost points of view, it offers certain significant advantages. For some applications (e.g., lighting and operating motors) it is indispensable. It carries with it, however, two disadvantages— relative unreliability and operating cost penalties.

Both the National Electric Reliability Council [18] and the Federal Power Commission [19] are forecasting serious nationwide electricity shortages in the early 1980s. They question the ability of the electricity generating system to ensure reliable supplies of power to their customers, particularly in industry. Such electricity shortages are not unprecedented. In 1973, the Pacific Northwest faced drought conditions that resulted in reduced electric supply and curtailments of industrial activity [19]. That condition was repeated in 1977, as the Bonneville Power Administration sought voluntary reductions in activity from all major industries within the region [20]. Other regions have been faced with shortages of varying severity and will, again,

face similar problems in the future. While entire reports have
been devoted to causes of this shortage, it is sufficient to
observe here that electricity is a difficult source of energy for
industry to count on for all of its needs.

Price is the second major obstacle in the extensive use of
electricity. Already, many industrial customers are paying from
$5 to $10/MM Btu (compared to $0.40 - $1.00/MM Btu for coal). It
is the policy of many state governments, and of the Federal government, to eliminate declining rates for large purchases. Further, peak load pricing on top of flat rate pricing can only esca-
late the cost of this energy source to still higher levels [3].
Given such a price escalation trend, it is difficult to foresee
electricity becoming a universally useful industrial fuel.

2. Synthetic Fuels from Coal. Because the remaining chapters of
this book provide a detailed examination of the options available
when using synthetic fuels, comments here will be abbreviated. It
is sufficient to observe that the resource base is large (see
section on coal) and the feed coal can become available. The
technology exists and its fuel products can feed existing boilers.

Synthetic fuels must find their applications within the capital cost differential of $15 - $60/lb steam/hr that exists between
boilers based on liquid as gaseous fuels and boilers based on
coal [7]. Or they must find a market in the capital/operating
trade-offs between coal combustion and electricity utilization.
The systems for gasification and liquefaction, the nature and
utility of the fuels produced, and the limits or constraints on
coal conversion will be examined in detail to see where, and how,
these fuels fit within the parameters established.

2.3 COMPARATIVE ANALYSIS OF FUEL OPTIONS

Certain comparative criteria can be employed to assess the
variety of fuel options available to industries and utilities.
These include resource reliability, resource abundance, regional
resource restrictions, current availability, ability to provide
space and process heat, and compatibility with existing (boiler)
investments. These comparisons are presented in Table 2.3.

As Table 2.3 suggests, each alternative has some applicability as utilities and manufacturers seek to convert their systems
from domestic oil and gas to substitute fuels. No single alternative offers a universal answer to the problem. It is within this
context—one alternative among many—that the synthetic fuels
options must be evaluated.

TABLE 2.3 Comparison of Alternative Fuels

Fuel	Resource Abundance		Resource Reliability		Regional Resource Limitation		Use Technology Currently Available		Supplies Space Heat		Supplies Process Heat		Compatible with Existing O & G boilers and Kilns	
	Yes	No	Yes	No	Yes	No	Yes	No	Yes	No	Yes	No	Yes	No
Imported Oil		X		X		X	X		X		X		X	
Coal	X		X			X	X		X		X			X
Oil Shale	X		X			X		X	N/A^a		N/A		N/A	
Residues		X	X			X	X		X		X			X
Nuclear		X	X			X	X		N/A		N/A			X
Geothermal		X	X		X		X		X		X			X
Solar	X		X			X	X		X			X		X
Electricity	X			X		X	X		X		X			X
Synthetic Fuels	X		X			X	X		X		X		X	

^{a}N/A = not applicable.

REFERENCES

1 Personal communication from Dr. James Boyd, President of Materials Associates, Inc. to W. Kenneth Davis, Vice President of the Bechtel Companies, January 28, 1977.

2 W. H. Axtman, "Shift in Boiler Fuel Usage Patterns," American Boiler Manufacturers Association, 1977.

3 Executive Office of the President, The National Energy Plan, U.S. Government Printing Office, April 1977, pp. 14-15.

4 Nuclear Energy Policy Study Group, Nuclear Power: Issues and Choices, Ballinger Publishing Co., Cambridge, Mass., 1977, pp. 94-96.

5 "Coal Report," a report by the Coal Sub-Panel to the Committee on Nuclear and Alternative Energy Systems of the National Academy of Sciences/National Academy of Engineering, April 1977.

6 David A. Tillman, "Converting to Coal Is More Complex Than Just Changing Fuel Suppliers," Area Development, June 1976, p. 14.

7 Ginger Prichard, "Shift to Coal May Snag on Boiler Supply," Energy User News, April 11, 1977, pp. 1, 9.

8 William Culbertson and Janet K. Pitman, "Oil Shale," in United States Mineral Resources, Donald A. Brobst and Walden P. Pratt (Eds.), Washington, D.C.: U.S. Geological Survey Professional Paper 820, 1973, pp. 497-504.

9 Thomas H. Maugh, II, "Oil Shale: Prospects on the Upswing ... Again," Science, Vol. 198, No. 4321, December 9, 1977, pp. 1023-1027.

10 David A. Tillman, "Combustible Renewable Resources," Chem Tech, Vol. 7, No. 10, October 1977, pp. 611-615.

11 Larry L. Anderson, "A Wealth of Waste, A Shortage of Energy," in Fuels from Waste, New York: Academic Press, 1977, pp. 1-16.

12 George D. Voss, "Industrial Wood Energy Conversion," in "Fuels and Energy from Renewable Resources, David A. Tillman, Kyosti V. Sarkanen, and Larry L. Anderson (Eds.), New York: Academic Press, 1977, pp. 125-140.

13 James Boyd and L. T. Silver, "United States Uranium Position," Prepared for the ASME-IEEE Joint Power Conference, Long Beach, Calif., September 18-21, 1977.

14 C. J. P. Muffler, "Geothermal Resources," in <u>United States</u>
    <u>Mineral Resources</u>, Donald A. Brobst and Walden P. Pratt
    (Eds.), Washington, D.C.: U.S. Geological Survey Profes-
    sional Paper 820, 1973, pp. 251-262.

15 Beverly A. Hall, <u>Materials Problems Associated with the</u>
    <u>Development of Geothermal Energy Resources</u>, U.S. Bureau of
    Mines, May 1975.

16 "Sun Heating and Cooling Set for $3M Industrial Park,"
    <u>Energy User News</u>, November 8, 1976, p. 15.

17 <u>A National Plan for Energy Research, Development and Demon-</u>
    <u>stration: Creating Choices for the Future</u>, Vol. 1 (ERDA-76),
    U.S. Government Printing Office, 1976, p. 62.

18 <u>6th Annual Review of Overall Reliability and Adequacy of</u>
    <u>the North American Bulk Power Systems</u>, National Electric
    Reliability Council, July 1976.

19 Advisory Committee Report, <u>The Adequacy of Future Electric</u>
    <u>Power Supply</u>: Problems and Policies, Federal Power Commis-
    sion, March 1976.

20 "17 Bonneville Users Face 25% Power Cut," <u>Energy User News</u>,
    February 14, 1977, p. 18.

CHAPTER 3

THE DEVELOPMENT
OF SYNTHETIC FUELS

Often coal is called the fuel of industrialization. It was
our primary energy source during the late nineteenth century and
early twentieth century, the years when the structure of our
economy became well established. What is less well remembered,
however, is that a conversion process—the manufacture of coke—
spurred coal's ascendancy. The historical progression of coal
conversion then becomes a useful analytical tool in developing an
assessment of this energy option. Such a review established a
base for the discussion of the principles and basic techniques
available for fuel synthesis. The availability and suitability
of various U.S. coals can then be considered.

3.1 HISTORICAL DEVELOPMENT OF COAL CONVERSION

Prior to the eighteenth century, wood was the dominant fuel
throughout the world. While much evidence suggests that coal
mining had been practiced since the twelfth century, that mineral
fuel had made little impact on industry. Wood and charcoal
fueled the furnaces and smelters of Europe's metallurgical industry. Centuries of this energy consumption pattern led to crisis
in fuel availability caused by deforestation. The iron and steel
industry of Great Britain, concentrated in the Forest of Dean and
in Weald, for example, faced possible extinction due to severe
energy shortages. The entire nation was facing a severe energy
crisis due to the overcutting of wood resources. In 1709, however, Abraham Darby achieved the production of coke from coal,
and successfully produced iron in his Coalbrookdale blast furnace
with coke. This development ultimately shifted the location of
the English iron and steel industry to the Birmingham area, but
30

provided for its salvation [1]. From this auspicious beginning,
coal conversion grew. Gasification and liquefaction processes
emerged to make coal, potentially, a most versatile energy source.

3.1.1 The Development of Coal Gasification

Coal gasification followed on the heels of coke production
in a series of related developments. The first gas producers
appeared much like coke ovens. Coal was charged inside the vessel.
More coal, outside the vessel, was burned. In this basic pyrolysis scheme, methane and other light hydrocarbons were driven off.
These volatiles with a heat content of 550 to 650 Btu/cf, were
the coal gas sold and burned. Some 70 percent of the initial
charge remained as solid fuel [2].

Technological improvements were made continuously after the
first coal gas plants were installed. The basic improvement was
to add the gasification step to the distillation or pyrolysis
process. Air or oxygen was blown through the heated coal to
achieve the following reaction: $C + O_2 \rightarrow CO_2$. When the reactor
vessel achieved the necessary temperature, a reduction reaction,
$CO_2 + C \rightarrow 2\ CO$, occurred. Subsequently, steam would be introduced to obtain the water-gas shift reaction, $C + H_2O\ CO + H_2$.
Synthesis gas, mixtures of hydrogen and carbon monoxide were
formed in this manner [2].

Unlike the distillation gas, with a heat content of 550 to
650 Btu/cf, the synthesis gas offered 300 Btu/cf. Yields, however, were far greater. Producer gas systems, continuous air-
blown gasifiers yielding a fuel with 110 to 180 Btu/scf were also
introduced. These were limited to industrial applications due to
the prohibitive expense of transporting the nitrogen-laden, low
energy content fuel over any appreciable distances.

Gas manufactured principally from coal made significant contributions to the U.S. economy. During the 1920s, some 20,000
gasifiers supplied energy to the U.S. economy. These were both
utility and industrial plants. In 1933 that gas manufactured for
sale by utilities contributed 1.82 quadrillion Btu (quads) to the
U.S. economy. That contribution increased to 1.99 quads at the
start of World War II and to a peak of 2.68 quads in 1949 [3].

While town gas was serving residential and commercial users,
industrial producer gas reactors emerged to provide manufacturers
with low Btu gas. The Wellman reactor was invented, for example,
in 1896. During the 1920s Wellman producer gas systems commanded
50 percent of the industrial market. In 1948 during the decline
of coal gasification, the Galusha agitator was added to the system
in order to make gasification of caking coals possible. Like the
Winkler, Lurgi, and Koppers-Totzek units, the Wellman-Galusha
system is available commercially today. Numerous Lurgi, Kopper-

Totzek, Wellman-Galusha, and Winkler gasifiers are operated around the world today. They are employed, principally, to produce feedstocks for liquid fuel and ammonia. Little producer gas remains in use. It is, however reemerging.

3.1.2 The Development of Liquid Fuels

Coal gasification was developed over three centuries of engineering, but liquefaction remains a twentieth century breakthrough. Processes were pioneered during the 1920s and 1930s in Germany. They are the basis for much of today's technology.

The work of Friedrich Bergius brought him the Nobel prize for chemistry in 1931. His direct hydrogenation of coal at elevated temperatures (806°F) and pressures (3,000 - 10,000 psig) led to the production of gasoline and aviation fuel [4]. At the same time Matthias Pier and co-workers found sulfur resistant coal hydrogenation catalysts that reduced the severity of the environment required for liquefaction while improving conversion efficiency.

The Luena works of Germany was built in 1931 before the Nazi regime took hold, as the world's first major coal liquefaction works. Its ultimate capacity was 650,000 tons/yr of coal. It was based upon the chemical engineering efforts of Bergius and Pier, and produced 15,000 bbl/day of light liquid fuels [5].

Slightly earlier Franz Fischer and Hans Tropsch developed indirect liquefaction—converting synthesis gas into low octane gasoline and other liquids [4]. This was utilized along with the hydrogenation liquids to fuel the German World War II effort. In that conflagration, synthetic liquids from coal provided 90 percent of the fuel required by the Luftwaffe.

After World II, interest was high to explore the German technology used in the war effort for coal liquefaction. The U.S. Bureau of Mines, at Louisiana, Missouri did demonstrate the production of light liquids from coal on this basis, and from 1949 to 1953, they produced 200 barrels of gasoline and other liquids per day in this research. Then the liquefaction research was dropped until 1960. In the Republic of South Africa, the state sponsored Sasol Corporation commercialized the production of light liquids from coal. While the first Sasol plant is of modest size, Sasol II, a $2.6 billion effort, is designed to produce 50,000 bbl/day of liquids from South African coal.

It can be observed, then, that the fundamental principles of coal conversion have been developed, demonstrated, and periodically commercialized. The development of coal conversion today represents a return to this earlier fuel supply system aided—in large measure—by infusions of new technology. Those principles

and the relationship of those technical concepts to American
coals and the U.S. economy merit more careful consideration here.

3.2   PROPERTIES OF COAL

Coal is a combustible rock of variable composition. Although
coal has been utilized for several hundred years the exact chemical nature and the origins of coal are not known. From physical
evidence of plant debris in many coals they are assumed to be
mostly composed of plant remains mixed with inorganic material
deposited with, or washed into, the deposits later. The process
by which such plant remains was converted to coal have included
decomposition, compaction, dehydration, and the action of heat
and pressure. This process is referred to generally as coalification and has been divided into a biochemical stage (diagenesis)
and a geochemical stage (metamorphasis). Because of relative
ages of associated deposits and composition coalification generally has been assumed to proceed from peat to lignite to bituminous coal to anthracite if anthracite is eventually formed. Geologists estimate that at least five to eight feet of plant material
were required to form one foot of coal. Because coal is a sedi-
mentary deposit it occurs ·in beds or seams that may extend for
many miles. These seams may be flat as in many areas of the Western United States or may be titled or vertical if they have been
folded by crustal movement of the earth. Although peat is of
recent age and anthracites are the oldest coals time alone seems
to be of little importance in coalification for coals older than
lignite.

3.2.1   Classification of Coals

Since coal is a material of variable composition many methods
have been developed to classify coal in order to define the properties that are useful in burning coal to produce heat. Coal may
be classified as to grade or rank. The grade of coal has to do
with how much foreign material (sulfur, mineral matter, etc.) are
present in the coal while rank has to do with carbon content,
heating value, and oxygen content of the coal. Many countries
who utilize coal have developed classification systems to differ-
entiate between different coals. These include Belgium, Germany,
France, Italy, The Netherlands, Poland, Great Britain, and the
United States. Even within these countries different classifications have been used. An international system has also been
developed for classifying coals by rank, but is not generally
used in the United States at this time. The most popular system
used in the United States, the ASTM (American Society for Testing

Materials), is given in Table 3.1. As one can see there is no
one criterion that differentiates between the various ranks
throughout the classification. Anthracites are differentiated
from bituminous coals by agglomeration characteristics and fixed
carbon. Bituminous coals are distinguished from subbituminous
coals by agglomerating properties and partially by heating value,
while subbituminous coals are distinguished from lignite by heating value.

Terms used in the ASTM classification that are used routinely
are defined as follows:

Fixed carbon: That fraction of coal not volatilized (or
burned off) from a sample heated to 950°C for seven
minutes.

Calorific value (or heating value): The amount of heat
given off by a sample burned in oxygen completely in a
Standard ASTM test.

Agglomerating: If the carbonized residue from heating
1 gram of coal to 950°C for seven minutes produces a
coherent button or residue showing coke structure, it is
considered to be agglomerating.

The classification of coals by rank, while not completely
consistent or descriptive of all of the important properties of
coal, is in wide use because it has been found useful in relation
to the use of coal. Most of this use has been combustion. The
classification by rank is less useful in describing properties
important in the production of synthetic fuels from coal. Indeed,
since the production of synthetic fuels is not a reality in most
of the world and the processes are not understood chemically, no
classification would be practical at this time. There are some
coal properties that can be related to their conversion to synthetic fuels. For example, anthracites are so high in carbon,
and the carbon is graphitic in nature so that liquefaction of
such coals is almost impossible and gasification is difficult.
On the other hand, lignites are high in oxygen and water content
thereby requiring more hydrogen to convert such coals to synthetic
fuels. In general, bituminous coals have a good balance between
volatile material (light material that can be vaporized from the
coal by heating) and relatively low moisture and oxygen. These
seem to be the best coals for liquefaction generally and usually
for gasification, too. There is significant progress in also
liquefying lignite (as the Germans did in World War II) as well
as gasification of lignite and even peat. Table 3.2 gives some
representative U.S. coals with their rank classification (from
Table 3.1) and important analytical data. Important compositional

factors that are not part of the ASTM (or other) classifications
of coal, but that are important in coal combustion and processing
include sulfur content, forms of sulfur, ash composition, and content and physical hardness and friability. Some of these are discussed in later chapters in connection with specific conversion
process and with environmental considerations.

3.3 PRINCIPLES OF COAL CONVERSION

Coal has experienced some problems in utilization for a variety of reasons. These include 1) it is a solid, 2) it contains
inorganic material that is not useful as a fuel, 3) it contains
some elements, the oxidation of which, produce environmentally
objectionable compounds (i.e., sulfur, mercury, nitrogen, etc.),
4) it is not of uniform composition. Conversion of coal to liquids or gases essentially eliminates these four disadvantages for
coal, although some separation and chemical reactions are
required. In order to convert coal to a liquid fuel or fuel gas,
the most important chemical change required is the addition of
hydrogen. The amount of addition determines the properties of
the synthetic fuel and the cost (in energy and dollars) of effec-
ting the conversion. A comparison of some representative materials illustrates this situation and is given in Table 3.3.
The conversion to high yields of methane, or substitute natural gas, is seen to be the most expensive process in terms of
hydrogen. This does not necessarily mean such conversion is the
most expensive or difficult. Process conditions, the catalyst
necessary for the reaction, the recovery of catalysts and technical problems such as feeding and product separation have major
significance, too.
In addition to adding hydrogen to the coal material in conversion to liquids or gases, elimination of inorganic matter and
reduction of sulfur, nitrogen, and oxygen are accomplished to a
greater or lesser extent, depending on the particular process
used. Although a very large number of processes have been attemp-
ted for coal conversion to synthetic fuels these are all of a few
basic types, all of which were pioneered by German scientists
before World War II.
In following chapters these processes are discussed along
with the characteristics of the products produced. The processes
for production of synthetic fuels and chemicals from coal may be
categorized in several ways, but the most simple is probably by
the type of products. Those to be discussed in following chapters
include the following:

TABLE 3.1 ASTM Classification of Coals by Rank

Class	Group	Fixed Carbon Limits (%) (daf basis)[a]	Calorific Value limits (Btu/lb) (Moist mineral matter- free basis)[b]	Agglomerating Character
I. Anthracite	1. Metaanthracite	98 (minimum)	—	Nonagglomerating
	2. Anthracite	92–98	—	Nonagglomerating
	3. Semianthracite	86–92	—	Nonagglomerating
II. Bituminous	1. Low volatile bituminous	78–86	—	Nonagglomerating
	2. Medium volatile bituminous	69–78	—	Nonagglomerating
	3. High volatile A bituminous	Less than 69	14,000 or greater	Nonagglomerating
	4. High volatile B bituminous	Less than 69	13,000–14,000	Nonagglomerating
	5. High volatile C bituminous	Less than 69	11,500–13,000	Nonagglomerating
			10,500–11,500	Agglomerating
III. Sub-bituminous	1. Subbituminous A	Less than 69	10,500–11,500	Nonagglomerating
	2. Subbituminous B	Less than 69	9,500–10,500	Nonagglomerating
	3. Subbituminous C	Less than 69	8,300– 9,500	Nonagglomerating

36

IV. Lignitic   1. Lignite A
                      (black lignite)   Less than 69   6,300- 8,300   Nonagglomerating
            2. Lignite B
                      (brown coal)   Less than 69   Less than 6,300   Nonagglomerating

[a]daf basis = dry mineral matter-free basis as determined by ASTM analysis.

[b]Moist mineral matter-free basis = mineral matter free basis by ASTM analysis, but with coal contining its inherent moisture.

TABLE 3.2 Some United States Coals Representative of the ASTM Classification by Rank[a]

Coal Rank (from Table 2.1)				Coal Analysis Bed Moisture Basis						Rank FC (dry mmf)	Rank Btu (moist mmf)
Class	Group	State	County	M	VM	FC	A	S	Btu		
I	1	Pennsylvania	Schuykill	4.5	1.7	84.1	9.7	0.77	12,745	99.2	14,280
I	2	Pennsylvania	Lackawanna	2.5	6.2	79.4	11.9	.60	12,925	94.1	14,880
I	3	Virginia	Montgomery	2.0	10.6	67.2	20.2	.62	11,925	88.7	15,340
II	1	W. Virginia	McDowell	1.0	16.6	77.3	5.1	.74	14,715	82.8	15,600
II	1	Pennsylvania	Cambria	1.3	17.5	70.9	10.3	1.68	13,800	81.3	15,595
II	2	Pennsylvania	Somerset	1.5	20.8	67.5	10.2	1.68	13,720	77.5	15,485
II	2	Pennsylvania	Indiana	1.5	23.4	64.9	10.2	2.00	13,800	74.5	15,580
II	3	Pennsylvania	Westmore-land	1.5	30.7	56.6	11.2	1.82	13,325	65.8	15,230
II	3	Kentucky	Pike	2.5	36.7	57.5	3.3	.70	14,480	61.3	15,040
II	3	Ohio	Belmont	3.6	40.0	47.3	9.1	4.00	12,850	55.4	14,380
II	4	Illinois	Williamson	5.8	36.2	46.3	11.7	2.70	11,910	57.3	13,710
II	4	Utah	Emery	5.2	38.2	50.2	6.4	.90	12,600	57.3	13,560

		State	County	M	VM	FC	A	S	Btu		
II	5	Illinois	Vermilion	12.2	38.8	40.0	9.0	3.20	11,340	51.8	12,630
III	1	Montana	Mussel-shell	14.1	32.2	46.7	7.0	.43	11,140	59.0	12,075
III	2	Wyoming	Sheridan	25.0	30.5	40.8	3.7	.30	9,345	57.5	9,745
III	3	Wyoming	Campbell	31.0	31.4	32.8	4.8	.55	8,320	51.5	8,790
IV	1	No. Dakota	Mercer	37.0	26.6	32.2	4.2	0.40	7,255	55.2	7,610

NOTE: Definition of coal rank is given in Table 2.1.

M = equilibrium moisture, %; VM = volatile matter, %; FC—fixed carbon, %; A = ash, %; S = sulfur, %; Btu = high heating value (Btu per pound). mmf = mineral matter free.

[a]Adopted from Van Nostrand's Scientific Encyclopedia (5th ed.), Van Nostrand and Reinhold Co., NY, p. 572, 1976.

TABLE 3.3   Hydrogen/Carbon Ratio and Hydrogen Content of Fuels

Fuel	Typical Atomic Ratio Hydrogen/ Carbon	Hydrogen Content (Weight %)
Methane	4.0	25
Iso octane (C_8H_{18})	2.25	15.8
Pennsylvania Crude Oil	1.98	15
Shale Oil	1.60	11.8
Texas Crude Oil	1.55	11.5
Tar Sand Bitumen	1.50	11.1
Coal Tar	1.33	10.0
Benzene (C_6H_6)	1.0	7.7
Bituminous Coal	0.80 – 1.0	5 – 6.6

1. Low Btu fuel gas—produced from coal by reaction with steam and air.
2. Medium Btu fuel gas ("synthesis gas")—produced from coal by reaction with steam and oxygen.
3. High Btu fuel gas (substitute natural gas)—This gas is produced by a three step process.

   a. Coal reacted with steam and oxygen to produce a medium Btu fuel gas consisting of mostly hydrogen and carbon monoxide.
   b. This "synthesis gas" is then "shifted," that is, altered to produce a mixture of H_2 and CO of the proper ratio.
   c. Methanation of the H_2/CO mixture to produce a fuel gas high in methane.

4. Heavy liquid fuels—produced from coal by pyrolysis, solvent refining, hydropyrolysis liquid phase hydrogenation, or by synthesis gas production followed by catalytic reaction of the synthesis gas to higher hydrocarbons.
5. Light liquid fuels and chemicals—produced from coal by one of the above processes with, in some cases, additional hydrogenation (vapor phase) or other catalytic or

high temperature reaction. (Such processing could
lead to mixtures of liquids with no predominant compound produce or to high yields of methanol or other
chemicals.)

3.4 THERMODYNAMICS OF COAL CONVERSION

The thermodynamics of a process or reaction are important
because the thermodynamic information about the change involved
indicates whether the undertaking of such a process or reaction
can be practical. It cannot give a definite "yes" to the prac-
ticality of a change, but it tells us when the answer is a definite "no." Thermodynamic information is that information about a
process concerning the spontaneity of a reaction and under what
conditions (temperature, pressure, etc.) the reaction is spontan-
eous, that is, conditions that favor formation of the products of
the reaction. If the thermodynamics are favorable under certain
conditions then kinetic information obtained under such conditions indicate if products can be obtained at a practical rate.

An example of a favorable situation (thermodynamically) that
proves to be difficult in practice is the reaction of coal with
steam to produce methane:

$$coal + 2H_2O_{(g)} \rightarrow CH_4 + CO_2 \tag{1}$$

At ambient conditions (25°C) the free energy change for this reaction amounts to about zero (assuming coal to be carbon). This
shows that the equilibrium constant would allow the reaction as
written to form products about as well as reactants. However, no
large scale processes to carry out this reaction have been developed because a suitable catalyst has not been found, the rate
being the limiting factor. As pointed out by Mills the reaction
of coal with steam has been catalzyed to some degree using alkaline catalysts such as potassium or sodium [6]. Still no process
utilizing this reaction has been implemented. The accomplishment
of the direct production of methane from coal would be a significant breakthrough. This will only be effected by improvement of
the reaction rate (i.e., suitable application of catalysts).

In the process of the hydrogenation of coal to produce synthesis gas some removal of heteroatoms (oxygen, sulfur and nitrogen) is accomplished.

This is desirable, provided not too much hydrogen is used,
since the heating value of the fuel is improved and the removal
makes resulting products more acceptable environmentally. Sulfur
is especially important to remove as there are specific regulations on the permissible levels permitted in fuels. Additionally
sulfur oxides in coal or other fuels corrode boilers and present

health problems. The conversion of coal by hydrogenation, pyrolysis or other methods usually results in significant reductions in
sulfur content, mostly by attacking the inorganic sulfur forms.
Organic sulfur, especially thiophenic sulfur is more difficult
(and expensive) to remove and nitrogen is even more refractory.
Fortunately, nitrogen is seldom present at high levels in coal
derived products.

In the reaction of coal with steam, or in other stages of the
gasification reactions, sulfur and oxygen are removed from the
coal as H_2S and H_2O. Hydrogen is used in the formation of these
products and the extent of removal has to be weighed against the
improvement in the environmental quality of the products.

One additional point should be made as to the engineering
problems associated with coal conversion; although reaction (1)
for direct methane production from coal is thermally neutral (not
much heat is absorbed or given off by the reaction). This is not
true for most coal conversion processes. For example, in the
high pressure coal liquefaction pilot plant operated at Louisiana,
Missouri (1949 to 1953) one of the most serious engineering problems was control of stable reaction conditions by the removal of
heat. While gasification reactions (coal + $H_2O \rightarrow CO$, CO_2, H_2 and
CH_4) are endothermic, methanation reactions ($3H_2 + CO \rightarrow CH_4 + H_2O$)
are highly exothermic. Thus, both liquefaction and gasification
of coal (to high Btu gas) involve engineering challenges to maintain temperature control.

3.5 CONCLUSION

Coal, the fuel of industrialization, has been highly utilized as a combustible fuel. Its use as a feedstock for producing
liquid and gaseous fuels has been less but significant. In the
present world environment of decreasing availability of petroleum
and natural gas, especially in the United States and Europe, coal
is a natural candidate for the raw material for liquids and gases.
The reasons for this are the abundance of the reserves in the
United States and Europe and the history of coal conversion summarized in this chapter. The history shows that technological
feasibility has been demonstrated. Surprisingly, major technological advances beyond that achieved by the Germans in the 1940s
have not materialized. This can be attributed to the lack of
incentive to produce synthetic fuels from coal resulting from the
availability of cheap clean fuels, oil, and natural gas. This
lack has delayed, until recently, research and development on the
chemical structure and reactivity of coal as well as the technology of conversion to more convenient energy forms.

One could argue that coal should be used as a solid rather than converting it to a liquid or gas since such conversion extracts part of the energy of coal. Undoubtedly some new methods for utilizing coal as a solid will be developed and present uses will be made more efficient. The production of synthetic fuels from coal will still have justification because of the convenience of using liquids and gases and our addiction to energy use in these forms due to the ways in which transportation and domestic and commercial systems have been developed. In the process of conversion coal is also made more environmentally acceptable by the reduction of inorganic material and sulfur in the synthetic fuel.

As petroleum and natural gas become more expensive in the United States, both in terms of price and in surrender of national security, coal will assume a more important role in the total energy supply. Synthetic fuels must at least be a real possibility, with technology demonstrated. More than likely a synthetic fuels industry will provide fuels in terms compatible with current and future utilization patterns as newer alternate energy sources are developed. Although a nonrenewable energy source, coal is still our most abundant fossil fuel and must carry the load until nonfossil sources are developed.

REFERENCES

1 G. M. Trevelyan, History of England, Vol. III, Garden City, NY: Doubleday and Company, 1953 (Anchor ed.).

2 Harry Perry, "The Gasification of Coal," Sci. Amer., Vol. 230, No. 3, March 1974, pp. 19–25.

3 Information Please Almanac, 1964, New York: Simon & Schuster, 1963, p. 589 (data from American Gas Association).

4 Martin A. Elliott, Howard R. Batchelder, Harlan W. Nelson, and G. Alex Mills, Fuel Chemistry—A Mid-Century Perspective, Columbus, Ohio: Battelle Memorial Institute, Nov. 1974.

5 E. E. Donath and Maria Hoering, "Early Coal Hydrogenation Catalysis," Fuel Processing Technology, Vol. 1, No. 1, August 1977.

6 G. Alex Mills, "Alternate Fuels from Coal," Chem Tech, July 1977, pp. 418–423.

CHAPTER 4

LOW BTU GAS FROM COAL

4.1 INTRODUCTION

Coal not only comprises the major portion of the fossil
energy resources of the United States, it is more extensively
used each year as the ability to obtain liquid and gaseous fuels
diminish. The largest single use of coal is for the generation
of electric power. However, the direct combustion of coal for
power production without stack gas cleanup has environmental consequences that the public objects to, particularly in the case of
high sulfur coals. Gasification of coal to low Btu gas is one
logical alternative to direct combustion combined with stack gas
scrubbing.

Coal gasification followed by gas combustion is inherently
less efficient than direct combustion (without regard to stack
gas cleanup). Taking the course of gasification indicates that
an environmental/economic choice has been made in favor of a
cleaner burning fuel with less air pollution. By choosing this
course one must be aware that more coal must be mined, more land
must be disturbed, and more water used to obtain the environmental
advantages of a cleaner fuel at the combustion site.

Commercial processes are available for low Btu gasification
of coal by Lurgi, Winkler, and Wellman-Galusha processes [1-4].
Variations of these gasification systems and others under development will undoubtedly continue to be used in new power plants
using combined cycle systems (gas turbines plus steam turbines).
The advantages of such systems have been brought about by tight-
ening of air quality standards and the combination of low supply
(relative to demand) and increased cost of natural gas and oil.

44

There is also incentive to utilize low Btu gas in conventional steam cycle power plants despite the somewhat higher capital cost for the same reasons given above.

Retrofitting of power plants to utilize low Btu gas is indicated where facilities exist for the handling and preparation of coal. Installations designed and built especially for using only natural gas or oil usually cannot practically change to low Btu gas. The most severe conversion problems are the absence of coal handling facilities and the stacks for combustion gases, that are generally not of sufficient capacity for low Btu gas combustion.

Gasification of coal to low Btu gas with subsequent combustion should be recognized as a viable alternative to other gaseous fuels and comparisons should be made based on the heating value per unit volume of fuel-air mixture of combustion products.

Transportation of low Btu gas outside of an industrial park or power complex is not practical. There is also some lower capacity limit for a practical low Btu gasification facility. However, in addition to use as "power gas" there are undoubtedly markets such as industrial boilers and furnaces and agriculture for low Btu gas that will continue to expand in future years.

4.2 PRODUCTION SYSTEMS

Renewal of interest in coal gasification has resulted in extensive evaluation of all gasifier types and the effects of operating conditions, as well as feed and product composition. Several authors have reviewed the past and present methods that have been utilized in the gasification of coal. As many as forty different processes have been suggested for producing low Btu or medium Btu fuel gas [4]. These methods can be categorized in several ways, one way being a description in terms of the types of gasifiers used. In this and the following chapters on gasification the following gasifier types are considered:

1. Fixed or moving-bed (slagging and dry).
2. Entrained flow (slagging and dry).
3. Fluidized bed (agglomerating and dry).
4. Molten salt.

Underground gasification can produce a low Btu fuel gas and is discussed briefly.

4.2.1 General Steps in Coal Gasification

Several methods of producing a low Btu gas (fuel gas with higher heating value less than 250 Btu/standard cubic foot) are described. In all of these methods some common basic steps are included although the inclusion of all of these basic steps is not necessary for every gasification process depending on the properties of the feed coal. The following steps are usually required or take place in gasification (in many cases the regions where these steps take place are not physically separate).

1. Pretreatment. Most bituminous coals have some caking proper-ties that cause difficulty in gasification if not treated to eliminate this property. The usual procedure is to perform a mild oxidation to eliminate caking. With such oxidation there is a loss of heating value in the range of 10 to 15 percent of that of the original coal.

2. Gasification. This can be divided into several steps that take place after the coal is introduced into the gasifier. The usual sequence is as follows.

   a. Drying and Heating. "Free-water" is removed in this step along with some "bound-water" (that is held in the submicroscopic pores of the coal). During this step the temperature increases to a value above the boiling point of water.
   b. Devolatilization (or distillation). Coals contain some materials that can be gasified easily, or evolved from the main coal material as gases with heating to tempera-tures below the softening temperature of the coal. The gases evolved are hydrocarbons, water vapor, and other light gases.
   c. Chemical Reaction. When the hot coal material is exposed to air and steam at ignition temperatures several reac-tions may take place.

$$C + O_2 \rightarrow CO_2 \qquad (1)$$

$$H_2 + 1/2\ O_2 \rightarrow H_2O \qquad (2)$$

$$C + 1/2\ O_2 \rightarrow CO \qquad (3)$$

$$CO_2 + C \rightarrow 2CO \qquad (4)$$

$$CO + 1/2\ O_2 \rightarrow CO_2 \qquad (5)$$

$$C + H_2O \rightarrow CO + H_2 \tag{6}$$

$$C + 2H_2O \rightarrow CO_2 + 2H_2 \tag{7}$$

$$C + 2H_2 \rightarrow CH_4 \tag{8}$$

$$C_nH_m \rightarrow \frac{m}{4}CH_4 + \frac{n-m}{4} C \tag{9}$$
(from volatile matter)

$$CO + H_2O \rightarrow CO_2 = H_2 \tag{10}$$

Reactions (1) and (2) are combustion reactions that release heat and proceed to completion in the presence of sufficient oxygen. Such sufficiency is not maintained in gasifiers since this would lead to complete combustion of the coal. However, these reactions are important in supplying heat for some of the other reactions that are endothermic [i.e., (6) and (7)]. As the coal moves through the gasifier some reactions approach equilibrium before the product gases leave the reaction zone. Reactions (4) and (6) never reach their equilibrium values since this would require complete reaction of steam by (7) which approaches equilibrium at gasification conditions. Reaction (7) does not proceed to CO_2 and H_2 below 980°F, while reaction (8), also involving carbon, mostly proceeds to CH_4 below this temperature.

In general, the use of steam in low Btu gas production tends to lower the temperature of the reaction zones by endothermic reactions such as (6) and (7). If these reactions proceed extensively and therefore reduce the steam partial pressure, the shifting of carbon monoxide [by reaction (10)] approaches equilibrium. At even higher steam decomposition reaction (10) can proceed in a backward direction (to produce CO and H_2O), provided sufficient CO_2 is present. In low Btu gas producers steam decomposition often reaches 90 percent. Reaction (10) is very important in all gasification systems and is discussed further in Chapters 4 and 5.

A complete treatment of the gasification reactions and their thermodynamic and kinetic effects under gasifier operating conditions is beyond the scope of the text. Details on the reactions and their significance to gasification are available elsewhere [3], [5-7].

Table 4.1 lists the gasifiers considered candidates for development or commercialization according to the above gasifier types [8-10]. This table includes systems for producing low Btu gas only, although many of those listed can and are being developed for either medium or high Btu fuel gas production. Heating values and product compositions are representative and may change

TABLE 4.1 Low Btu Gasifiers and Characteristics

Gasifier Type and Licensor	Feed Requirements	Operating Conditions T(°F)	Operating Conditions P(psig)	HHV (Btu/ft³)
Fixed (or moving) Bed				
USERDA-MERC (ERDA-Morgantown, Energy Research Center)	-1/2" (50%)	2450	atm -285	150
Lurgi (American Lurgi Corporation)	1/8" × 1-1/2"	1200-1500	350-450	180
Gegas (General Electric Company)	-1/8"	-	290	<300
Wellman-Galusha (McDowell-Wellman Engineering)	1"-2" (bit. coal)	~2400	atm	168
Woodall/Duckham/Gas Integrale (Woodall-Duckham [U.S.A.] Ltd.)	1/4"-1-1/2" (FSW<2.5)	2200	atm	175
Entrained Flow				
Babcock & Wilcox	-200 mesh (70-90%)	3400	atm -300	102
Combustion Engineering	-200 mesh (70%)	3000	atm	127
2-Stage, Slagging (Foster-Wheeler)	-200 mesh (70%) <2% moisture	2500-2800	350	177
Slagging, Vortex Chamber, (Ruhrgas, A.G.)				100
BCR Low Btu (Bituminous Coal Research)	+200 mesh (40%) -200 mesh (60%)	2100	235	160

Synthane (ERDA-PERC)	-20 mesh	1800	1000	165
U-Gas Institute of Gas Technology	0" × 1/4"	1900	50–350	154
Westinghouse	0" × 1/4"	2100	130–200	135
Molten Salt				
Rummel (Union Rheinische Braunkohlen Kraftstoff A.G.C. Otto and Company)		>2100		110
AL Molten Salt (Atomics International, Division of Rockwell, International)	-1/4"	~1800	up to 280	158
Kellogg Molten Salt (M.W. Kellogg)	-12 mesh			

(Continued)

TABLE 4.1 (Continued)

Gasifier Type and Licensor	Gas Composition (Dry%)					
	CO	CO$_2$	H$_2$	Natural Gas	CH$_4$	Other HC's
Fixed (or Moving Bed)						
USERDA-MERC (ERDA-Morgantown, Energy Research Center)	16	12.6	19	48.6	3.8	—
Lurgi (American Lurgi Corporation)	13.3	13.3	19.6	48.3	5.5	—
Gegas (General Electric Company)	—	—	—	—	—	—
Wellman-Galusha (McDowell-Wellman Engineering)	b28.6 a27.1	3.4 5.0	15.0 16.6	50.3 50.8	15.1 .5	— —
Woodall/Duckham/Gas Integrale (Woodall-Duckham [U.S.A.] Ltd.)	28.3	4.5	17	47.2	2.7	—
Entrained Flow						
Babcock & Wilcox	23.3	4.5	8.4	63.7	—	—
Combustion Engineering	22.1	7	17	53.3	.03	—
2-Stage, Slagging (Foster-Wheeler)	29.3	3.3	14.5	48.7	3.5	—
Slagging, Vortex Chamber, (Ruhrgas, A.G.)	22.8	5.1	8.0	64.1	—	—

BCR Low Btu (Bituminous Coal Research)	25.7	5.2	23.4	45.7	–	–
Synthane (ERDA-PERC)	10.1	17.9	21.5	44.2	6.3	–
U-Gas Institute of Gas Technology	19.6	9.9	17.5	49.6	3.4	–
Westinghouse	19.2	9.3	14.4	54.4	2.7	–
Molten Salt						
Rummel (Union Rheinische Braunkohlen Kraftstoff A.G.C. Otto and Company)						
AL Molten Salt (Atomics International, Division of Rockwell, International)	29.7	3.5	13.2	49.4	1.5	–
Kellogg Molten Salt (M. W. Kellogg)						

b = bituminous coal
a = anthracite

with variaton in feed coal and/or operating conditions. The
various processes for production of low or medium Btu gases cover
a wide range including some that are available commercially and
several that have been only tested to a limited extent in bench
scale experiments. Those of interest in this chapter for low Btu
gas all use air (or air and steam) resulting in a high N_2 content
(37 to 63 percent) in the product fuel gas. Fuel gases lower in
N_2 (obtained when O_2 is used instead of air) contain relatively
higher concentrations of CO and especially H_2 and are considered
in the following chapter since their heating values are near or
above 300 Btu scf.

4.2.2 Systems Available for Low Btu Gas Production

General characteristics and descriptions of the gasifier
systems are given for those that produce low Btu gas. Although
in Table 4.1 many systems seem to be equivalent or nearly so,
there are wide differences in experience with these systems.
Often limited experience at higher capacities and/or with continuous operation make the claims of a particular system suspect relative to another system that has been operated continuously or at
pilot or full scale capacities.

Most of the gasifier systems are claimed by licensor and/or
developers to handle all types of coal (caking and noncaking).
In some cases these same systems are known to experience operating problems with caking coals. Specific problems will be discussed.

1. **Fixed Bed Gasifier Systems.** In fixed or moving bed gasifiers
coal is usually fed at the top of the gasifier and moves downward
through the gasifier while steam and air are injected into the
gasifier chamber at the bottom after which these gases pass upward through the coal bed. At the bottom of the bed carbon
reacts with oxygen to form carbon dioxide in a zone that is quite
thin. Above this zone the endothermic steam carbon reactions
[reactions (6) and (7)] takes place forming carbon monoxide and
hydrogen. As the hot gases from these reactions flow upward
through the bed carbon, monoxide is formed by the endothermic
reaction of carbon dioxide with carbon [reaction (3)]. The water
gas shift reaction [reaction (10)] takes place through the bed.
This means that as coal is injected at the top of the reactor it
first undergoes drying from contact with the hot product gases,
then devolatilization as the temperature rises, and finally reaction with oxygen in the air supplied and with steam giving off
gases. The volatile products from the devolatilization of the
coal crack to form hydrogen, methane, and higher hydrocarbons.

The residual material from a low Btu gasifier that does not react is discharged at the bottom of the bed as the gases flow out the top of the gasification chamber. In the case of slagging fixed bed gasifiers the residual material is discharged as a liquid slag. The hot fuel gases leave the top of the bed usually at a temperature of around 1000°F (540°C). The most serious problems with fixed or moving bed gasifiers are related to clinkering and formation of large coke zones in the bed due to softening and caking of certain coal material, particularly bituminous coals. Two examples are given below to illustrate the operational advantages and disadvantages of fixed bed gasifiers.

    a. Lurgi—Historically the Lurgi gasifier developed by Lurgi Mineröltechnik GMBH has received one of the largest applications in industrial operations since it was introduced. This process uses dry coal and claims to be able to handle all types of coals. However, both slightly caking coals and those with pronounced caking properties are often unsatisfactory in Lurgi gasifiers. Efforts to gasify U.S. caking coals by Lurgi technology have been made [11, 12]. Mechanical methods are used to deal with the properties of the caking coals when these are treated. Crushed coal is required as a feed, usually having a size less than one inch.
    Lurgi gasifiers are water-cooled and refractory lined and operated at pressures up to 20 atmospheres. Essentials of the Lurgi are shown in Figure 4.1 along with some of the important reactions.
    Some of the important disadvantages of the Lurgi process are:

    i.   Byproduct phenols and tars are produced that require a
         hot tar-scrubbing tower.
    ii. The gasifiers are small (12 ft-diameter) and process
         less than 300 tons of coal per day.
    iii. Fairly large amounts of steam are necessary to operate
         the system and prevent agglomeration and clinkering
         (0.80 lbs. steam and 2.7 lbs. air per pound of coal
         is usual).

    b. Wellman-Galusha—These gasifiers have been manufactured commercially since before 1940 in several countries. Coke, anthracite and bituminous coal have all been processed [13].
    Wellman-Galusha gasifiers have been used commercially in several countries. There were some seven operating in industrial plants in the United States in 1977. Wellman-Galusha gasifiers are small and operate near atmospheric pressure. There are two

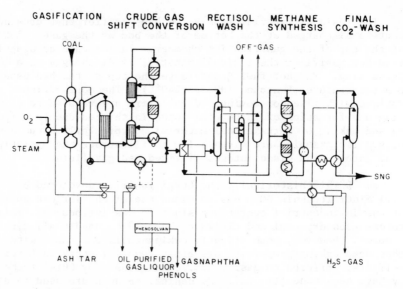

Figure 4.1. The Lurgi Process. This figure shows the complete Lurgi process to produce substitute natural gas from coal. If the oxygen supply is in the form of air, and if steps from crude gas shift conversion to final CO_2 wash are eliminated, the Lurgi can be used as a low Btu gas producer. It is the largest low Btu gas producer commercially available.

types, a standard type and an agitated type with the agitated type having about 25 percent higher capacity than the standard. Figure 4.2 shows the essential features of an agitated Wellman-Galusha gasifier. Another advantage of the agitated type is that it can handle caking bituminous coals. The Wellman-Galusha gasifier is water jacketed and the inner wall of the gasifier is steel plate rather than refractory as in the Lurgi gasifier. An agitator has a revolving horizontal arm that serves to retard channeling and maintain a uniform fuel bed in the gasifier. Coal is crushed to at least 9/16" in diamter for anthracite and less than 2" for bituminous coal. This crushed coal is fed into a coal bin and flows into the gasifier by gravity. Air and steam are usually blown into the gasifier by a fan with the operating temperature in the combustion zone at about 2400°F (1300°C). The gasifier has a diameter of about 10 feet and a capacity of 30 to 84 tons/day depending upon the type of feed material. The higher capacities are obtainable only with bituminous coals that are more reactive. Only a small amount of tar or phenol materials are produced even if bituminous coals are used in Wellman-Galusha

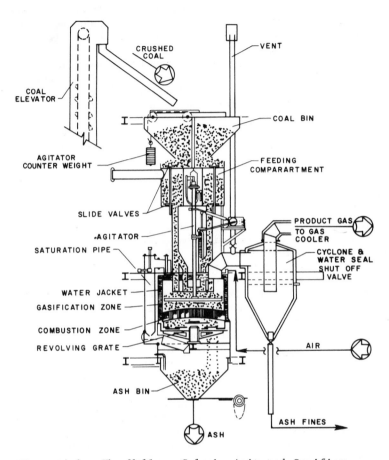

Figure 4.2. The Wellman–Galusha Agitated Gasifier.
Wellman – Galusha low ьtu gasifiers have been built in
sizes ranging from 45 to 85 tons/day of coal. It can
consume coke and anthracite as well as bituminous coal.
Several are in operation now in the United States.

gasifiers. These gasifiers have been used for about 35 years and all types of coal can be processed. For air blown operation approximately 0.4 pounds of steam and 3.5 pounds of air are required for each pound of coal processed.

Wellman-Galusha gasifiers have received recent interest by the fact that the Glen-Gery Brick Company of Reading, Pennsylvania, reactivated one unit in 1973 and began operation of a new unit at the beginning of 1975. These units (both in anthracite country) are located in Shoemakerville and Watsonville, Pennsylvania where they are used to supply fuel gas ($\sim$150 Btu/ft^3) to brick kilns. Although burners and air systems had to be redesigned to use the low Btu gas:air mixture (1.0:1.1 instead of the 1.0:10.6 used with natural gas) the mixture entering the brick kilns has a heating value of 79 Btu/ft^3, only 15 percent below the 93 Btu/ft^3 obtained previously with natural gas.

The proceeding descriptions have to do with fixed bed gasifiers that operate dry; that is, temperatures are below those required for slagging conditions. Slagging operation can be advantageous for these fixed bed gasifiers since it improves performance and eliminates the need for a grate or other mechanical devices for ash removal. For slagging conditions a high temperature, enough to melt the ash and maintain proper fluidity in the slag, is maintained so that tapping of this material can be done periodically. The particular temperature required depends on ash content and the composition of the particular ash.

Slagging fixed bed gasifiers have the advantage of higher thermal efficiencies that result from a lower steam requirement. The overall thermal efficiency advantage for slagging amounts to about 5 percent and results from operation with molar steam/air ratios in the range of 0.2-0.3 which is much nearer conditions of maximum thermal efficiency. (For dry fixed bed gasifiers the steam/air ratio molar ratios are usually 1.25 to 2.0).

Some examples of programs to investigate the advantages and the development of slagging gasifiers include the British Gas Corporation in England at Solihull in the 1960s, a 24 ton day pilot scale slagging gasifier that was operated in the 1950s and 1960s at the Grand Park Energy Research Center and the Thyssen-Galoczy. This latter gasifier was used on a commercial scale in Europe prior to 1950. The British Gas Corporation/Lurgi operated in Solihull is reportedly to have achieved the benefits of 4 to 7 times the capacity of a standard Lurgi gasifier and a higher thermal efficiency by 10 to 20 percent and also had the advantage of fewer pollutants compared with the standard Lurgi process. Agitated Wellman-Galusha gasifiers offer some interesting possibilities with their higher efficiencies over the standard type and have been more successfully used on a commercial scale than

Lurgi gasifiers when using caking coals. Wellman-Galusha units can be started up in only a few hours and can be operated at less than nominal capacity without adversely affecting gas quality. Another advantage is that the gasifiers can be banked for several days and the temperature maintained with only blowing of air for a few minutes each day to maintain temperature in the combustion zone.

2. Entrained Flow Gasifiers.

Entrained-flow gasifiers utilize the gas feed mixture (air/steam or oxygen/steam) to carry pulverized coal particles through the gasifier. The particle size of the coal and the gas velocity must be such that suspension is maintained. Considerable experience in this area has come from pulverized-coal-combustion systems. Since the concurrent streams in entrained flow gasifiers move at velocities higher than terminal velocities of the particles, ash and char particulates exit from the gasifier chambers with the product gases and a separator is required. Heat is also taken from the gasifiers rapidly as a result of the high flow rates necessitating heat recovery systems for efficient operation. Reaction temperatures and rates are high as compared with fixed bed systems. When the temperatures exceed slagging conditions relatively complete gasification can be achieved in a single pass. Dry systems usually employ char recycle systems to achieve high conversion.

Presently available technology includes Kopper-Totzek (K-T), although this suffers from low thermal efficiency and unavailability of large capacity units. K-T units were originally used to provide synthesis gas for ammonia manufacture. These gasifiers have good versatility, processing all ranks of coal and some have been used to gasify heavy oil. However K-T gasifiers are oxygen blown and produce medium Btu fuel gases (discussed in Chapter 4). Actually little success has been achieved for producing low Btu fuel gas from air blown entrained flow gasifiers. The concept has been pursued by Babcock and Wilcox, Combustion Engineering, and with the Szilka-Rozinek boiler. At present there seems to be little promise for the gasification of coal to low Btu gas in entrained flow gasifiers without a significant breakthrough in achieving the necessary high temperatures while substituting air for oxygen as a feed gas. Perhaps the use of "oxygen enriched air" would be possible in the entrained flow systems. In the United States there are several projects for developing entrained gasification technology at higher capacities. Table 4.2 lists these projects as of 1976 with capacities and estimated completion dates.

TABLE 4.2 Projects for Development of Entrained Flow-
Gasification

Method	Sponsors	Capacity (MW)	Date of Completion
Koppers-Totzek (pressurized)	Shell Oil Company Koppers Corporation, GMBH	15	1978
Combustion Engineering	EPRI, ERDA (DOE)	10/100	1978/1983
Babcock & Wilcox	EPRI	40	1980
Foster Wheeler	ERDA (DOE) and Electric Utility Industry	40	1981/1985
Texaco	Texaco, Inc.	15 TPD	1972

3. Fluidized Bed Gasifiers. Of the systems discussed so far,
fluidized-bed gasifiers are the least developed. The first commercial units were air-blown Winklers at the Leuna Works in Pre-
World War II Germany. These gasifiers were used to supply gas
for producing electric power to run gas engines used in the
ammonia synthesis. Five units were operated successfully for a
short time with air and lignite. In 1930 the Winkler units were
modified to blow oxygen and steam. This allowed reduction in
the bed temperature (from 1750°F or 800°C with oxygen) and also
in reduction of the temperature of the overhead space from 1000°
to 850°C (1832° to 1560°F). These changes eliminated the formation of sintered ash obstructions at the gas exits while maintaining a satisfactory utilization of carbon fines above the gasification bed. The five units put into operation in 1926 with air
continued operation with oxygen and steam until 1971 when they
were finally shut down. In 1972 Winklers were successfully operated at pressures above atmospheric (to ca. 22 psi) and plans
were formulated for development in pilot plants using U.S. coals
at pressures up to 200 psi. One problem with U.S. coals that was
not encountered with German lignites is that of handling highly
caking bituminous coals. Fluidized bed gasifiers can handle coals
of widely differing properties if agglomeration is minimized or
eliminated so that fluidization conditions are maintained.

With caking coals two methods have been used to avoid loss of fluidization by agglomeration. These are: 1) pretreatment of the caking coal—usually a mild oxidation with air or oxygen or 2) dilution of feed coal with nonagglomerating char. In fluidized bed gasification fine coal is kept fluidized by the velocity of the upward moving gas. As with entrained flow gasifiers considerable heat and fines are carried over in the fuel gas that exits from the gasifier. This disadvantage is offset by the suitability to continuous operation at high throughput. This leads to high space utilization rates (one pound of coal treated/unit time and reactor volume) and the favorable mass and heat transfer of fluidized bed gasifiers makes temperature control easier and more uniform.

Fluidized bed gasifiers may be operated under dry or slagging conditions. Dry fluidized bed gasifiers operate at temperatures below ash softening temperatures. Most char and ash particles are removed from the bottom by mechanical methods while some are elutricated with the fuel gas at the top and requires cyclones for separation. In slagging bed fluidized-bed gasifiers the temperature is above (but close to) the ash softening temperature so that ash particles contact one another and stick. Small clinkers result from this agglomeration of ash particles that grow until they are too heavy to remain fluidized whereupon they sink to the bottom of the bed and are removed. Table 4.3 lists the fluidized bed gasification systems under development for low Btu gas production.

4. Molten Salt Gasification. Molten substances can be used in gasification to transfer heat and catalyze the reactions. Substances that have been proposed as melt material include other substances besides salt, such as iron; other metals and coal ash. Gasification by molten media has the advantages of higher gas yields (due to catalytic action of the melt), particle size and caking properties of coals are unconsequential and sulfur can usually be retained in the melt for separate processing thus reducing the need for gas cleanup.

The major disadvantage for these gasification systems are corrosion problems associated with the high temperature melt. The production of low Btu gas from coal has been studied by both M. W. Kellogg and Atomics International (Rockwell International Corporation) and both systems use molten salt. Applied Technology Corporation has done experimental work on low Btu gasification of coal using molten iron in conjunction with a slag containing limestone (to reduce the sulfur in the coal); this is a variation of the Atgas process for SNG production.

TABLE 4.3. Fluidized-Bed Gasifiers, Low Btu Gas

Name	Developer	Conditions of Operation	1977 Stage of Development (capacities)
Dry Bed			
Winkler	Davy Powergas, Inc.	Atmospheric pressure; 1500–1800°F	Commercial
Synthane	ERDA–PERC	1000 psi; 1800°F	Pilot–72 TPD[a]
BCR–Low Btu	Bituminous Coal Research	Up to 235 psi; 600–2100°F	PEDU[b]–1.2 TPD
Slagging Bed			
U-Gas	Institute of Gas Technology	50–350 psi; 1900°F	PDU[c] – TPD
Westinghouse	Westinghouse Electric	130–200 psi; 1600–2100°F	PDU – 0.6 TPD (design – 1200 TPD)

[a]TPD: Tons/day.
[b]PEDU: Process and Equipment Development Unit.
[c]PDU: Process Development Unit.

a. Atomics International—The gasifier for the Atomics
International Al molten salt process is 3 ft (I.D.) × 10 ft high.
Coal, crushed to permit pneumatic transporting (usually - 1/4"),
and sodium carbonate are fed to the bottom of a molten salt bed
with air. Since the melt is maintained at ~1800°F the coal is
immediately pyrolyzed (and oxidized by the air feed) producing a
fuel gas of 130-160 Btu/ft^3. The gases escape out the top of the
melt while the sulfur and ash are retained.

The molten sodium carbonate performs several functions besides heat transfer in the gasification.

  i.   Acting as a catalyst that increases the specific
       reaction rates compared with noncatalytic systems.
  ii. Reaction with the sulfur in the coal.
  iii. Providing for dispersion of coal and air.
  iv. Reception of the inorganic material in the coal.

The molten salt is withdrawn continuously and treated to
obtain H_2S, coal ash, and the regenerated sodium carbonate which
is recycled to the gasifier. The 120 TPD pilot plant is programmed to produce 15.8 million scf/day of low Btu gas that has a
heating value of 130-160 Btu scf and results in a thermal efficiency of nearly 80 percent. Technical information obtained is
used for full scale plant design if the process appears to be
feasible and practical at commercial scale operation. The serious
problem of corrosion by the hot molten bath has been temporarily
solved by using a high purity alumina refractory material to contain the salt.

b. Kellogg Molten Salt—Kellogg has done considerable work
at small scale on the gasification of coal in a molten salt bath
similar to the Atomics International Al molten salt process. The
main thrust of the work has been directed toward medium Btu fuel
gas production using steam and oxygen and operating at higher
pressures (1200 psi). With the same system and air in place of
oxygen a low Btu gas is produced and the gasification can be carried out at lower pressures. In this process ash is separated
from the molten salt by filtration of a water solution of the
salt. Bicarbonate salt is precipitated from the solution and
then heated (calcined) to provide carbonate salt that can be
recycled to the gasifier. In the production of high Btu gas shift
conversion, methanation and dehydration are also necessary steps
in producing a pipeline quality gas.

The advantages of this process are similar to those cited
for the Al molten salt process. In addition the Kellogg process
includes operation at higher pressures with methanation of the
synthesis gas produced. However, a relatively large methanator

is required and the system is complicated by the calcination of bicarbonate.

c. Other Molten Systems—Several other molten bath systems have been studied at bench and larger scale. The two systems described here are those pursuing pilot plant programs at utility sites and therefore closest to commercial use. The Otto-Rummel (double shaft) and Rummel (single shaft) are alternative possibilities for molten bath gasification. The latter has been operated commercially in Germany and can be used to gasify char or liquid hydrocarbons as well as coal.

5. **Underground Gasification.** Gasification of coal in situ has, for many years, been looked upon by some as a possible solution to many problems of coal production and utilization [14, 15]. Such processing would consist of partial combustion of coal by injection into the coal seam of air-steam, oxygen-steam or air alone after an initial combustion has been initiated. Communication from the injection well to production well (or wells) would lead to gasification of the coal seam material. Studies have been carried out in Russia, England and the United States at various times since the 1930s. The most recent U.S. work field tests at Hanna, Wyoming, have produced low Btu gas (175 Btu/ft^3) [16].

Obvious advantages of underground gasification, if it could be developed successfully, include the following.

a. Utilization of coals not recoverable by conventional mining (too deep, too thin, low grade).
b. Significant reduction in personnel for production (no miners or transportation facilities for coal).
c. Reduction of environmental impacts of solid wastes disposal.
d. Production of a cleaner fuel.

There are still many unanswered questions on the economic and technical feasibility of underground gasification. Only extensive field or full scale tests show whether this technique can be used without pollution of ground water (many coal seams are aquifers) and without serious subsidence problems.

4.3   USE CHARACTERISTICS OF LOW BTU GAS

4.3.1   Quality of Low Btu Gas

Low Btu gas is primarily CO, H_2, and N_2. Minor quantities of H_2S, H_2O, CO_2, and CH_4 as well as trace amounts of other gases are also present. The quality of a particular low Btu gas product from a gasification system is only partially dependent on the relative amounts of the gaseous components listed above. The presence of char fines, ash particles, and liquid droplets must also be considered.

The heating value range of low Btu gas has been arbitrarily defined differently by several authors as 85-200 Btu/scf, less than 150 Btu/scf or less than 250 Btu/scf [17-19]. In this chapter it has been assumed that air blown gasification results in low Btu gas. Since H_2 and CO are the major fuel components (with higher heating values of 324 and 321 Btu/scf, respectively), and N_2 makes up about 1/2 of the total gas volume the heating values are generally below 200 Btu/scf. Significant amounts of CO_2 or O_2 or higher quantities of N_2 lower the heating value while the presence of any hydrocarbon gases increase it.

Gas produced from fixed or moving bed gasifiers usually have less suspended fines than that produced from entrained or fluidized bed units.

Gas quality, as determined by composition (heating value) and presence of suspended solids, depend primarily on the amount of N_2 and other inerts, the presence of H_2S, the amounts of any CH_4 and other hydrocarbons, and the method of production.

4.3.2   Behavior of Low Btu Gas

Low Btu gas usually involves combustion with air at ambient temperature. Depending on circumstances of production and use the gas may be supplied hot or cold; the air could also be preheated so these possibilities are discussed. Some comparisons are made with methane since it usually makes up over 90 percent of natural gas and low Btu gas applications are often made where natural gas has been utilized.

Table 4.4 gives flame temperatures for major components found in natural and low Btu gas. Reference [17] (and others) give additional information on the relationships between heating value and flame temperature and volume of flue gas produced. Probably more important information given in [17] is the actual combustion data on gas compositions that are produced from Lurgi, Wellman-Galusha, Winkler, Koppers-Totzek, and Babcock and Wilcox

TABLE 4.4 Maximum Flame Temperatures for Gases (Temperatures measured using Na line reversal method, combustible mixture initially at room temperature)

Fuel Gas Component	Oxidant	Combustible Fraction in a Stoichiometric Mixture	Combustible Fraction Used Total Mixture	Flame Temperature °F	°C
Hydrogen	air	0.296	0.316	3713	(2045)
CO	air	0.296	0.20	3002	(1650)
CO	air	0.296	0.25	3506	(1930)
CO	air	0.296	0.32	3812	(2100)
CO	air	0.296	0.37	3632	(2000)
Methane	air	0.0947	0.10	3407	(1875)
Ethane	air	0.0607	0.058	3443	(1895)
Propane	air	0.0403	0.0415	3497	(1925)
Power Gas 17 (160 Btu/scf)	air	–	–	3100	(1704)

gasifiers. This information as it applies to general low Btu gas combustion makes the following points:

1. For fuel gases having heating values below 200 Btu/scf more than 12,000 scf of flue gases are produced per 10^6 Btu in the fuel gas, with the amount of flue gas increasing with decrease in higher heating value.

2. As indicated in Table 4.5, low Btu gases require much less air for combustion (from 6 to 16 percent of that required for methane) and produce larger volumes of flue gases (on a Btu basis).

4.3.2 Environmental Aspects of Low Btu Gas Use

As shown by earlier sections of this chapter there are several systems that are available for production of low Btu gas from coal. When coals containing sulfur to any appreciable extent are gasified desulfurization is necessary. The total effect on the environment utilizing gasification and desulfurization is positive relative to conventional combustion. Reductions in contaminants to the environment should be significant. Expected reductions from direct use of coal are:

90 percent in sulfur dioxide

Nearly 100 percent in particulates

Significant reduction in NO_x (due to lower combustion flame temperatures)

One negative aspect of coal gasification that is significant in dry areas of the Western United States is that such conversion requires large quantities of water. The requirements vary somewhat depending on the gasification method, but conventional power plants with coal gasification require at least 15 percent more water than a power plant utilizing coal, even with SO_2 scrubbing. Combined cycle power plants with coal gasification require about the same amount of water as coal fired plants without stack gas scrubbing. These issues are treated more extensively, along with considerations of economic costs, in Chapter 9.

TABLE 4.5 Air Requirements and Flue Gases Produced by Low Btu Gas Combustion

Fuel Gas	Stoichiometric Air Required		Flue Gas Produced from Stoichiometric Mixture Combustion	
	SCF/SCF Fuel	SCF/10^6 Btu	SCF/SCF Fuel	SCF/10^6 Btu
Methane	9.52	9425	10.52	10,415
Lurgi (from Sub-bit. coal)	1.55	7990	2.35	12,115
5.1% CH_4 [a]				
0.2% C_2H_4				
0.4% C_2H_6				
17.5% CO				
23.3% H_2				
14.9% CO_2				
38.6% N_2				
(HHV = 194 Btu/scf)				
Wellman-Galusha (sub-bit. coal)	1.05	7780	1.88	13,925
2.7% CH_4				
15.6% CO				
17.7% H_2				
14.2% CO_2				
49.8% N_2				
(HHV = 135 Btu/scf)				

Winkler (lignite) 0.7% CH_4 22.7% CO 12.7% H_2 7.8% CO_2 56.1% N_2 [a] (HHV = 121 Btu/scf)	0.91	7520	1.73	14,295
Babcock & Wilcox (Sub-bit. coal) 18.5% CO 7.8% H_2 8.3% CO_2 65.4% N_2 (HHV = 85 Btu/scf)	0.626	7365	1.49	17,550

[a] Percentages of each component (volume basis) in the fuel gas.

REFERENCES

1 A. M. Squires, "Clean Fuels from Coal Gasification," Science, Vol. 184, No. 4132, pp. 340-346, 1974.

2 National Research Council, "Evaluation of Coal Gasification Technology," Part II (Low and Intermediate Btu Fuel Gases) 1973.

3 L. K. Mudge, G. F. Shiefelbien, C. T. Li, and R. H. Moore, "The Gasification of Coal," Battelle Pacific Northwest Laboratories, A Battelle Energy Program of Coal Report, July 1974.

4 I. Howard-Smith and G. J. Werner, "Coal Conversion Technology" Noyes Data Corporation, Park Ridge, NJ, 1976.

5 W. Gumz, Gas Producers and Blast Furnaces, New York: John Wiley and Sons, 1950.

6 H. H. Lowry (Ed.), Chemistry of Coal Utilization, Supplementary Volume, NAS-NRC, New York: John Wiley and Sons, Chap. 20, pp. 892-1022, 1963.

7 W. L. Lom, and A. F. Williams, Substitute Natural Gas, Applied Science Publ. Ltd., London, Chap. 5, pp. 80-94, 1976.

8 U.S. Dept. of Commerce (National Technical Information Service) prepared by Dravo Corporation, Handbook of Gasifiers and Gas Treatment Systems, prepared for the U.S. Energy Research and Development Administration, February 1976.

9 P. F. H. Rudolph, "The Lurgi Process Route Makes SNG from Coal," Oil Gas J., Vol. 71, No. 4, pp. 90-92, 1973.

10 J. D. Cooperman, D. Davis, W. Seymour, and W. L. Ruckes, Lurgi Process - Use for Complete Gasification of Coals with Steam and Oxygen Under Pressure, U.S. Dept. of Interior, U.S.B.M. Bulletin 498, 1951.

11 D. C. Elgen and H. R. Perks, "Results of Trials of American Coals in the Lurgi Pressure Gasification Plant at Westfield, Scotland," Proceedings of Sixth Synthetic Pipeline Gas Symposium, Oct. 1974.

12 R. E. Morgan, J. W. Eckard, J. Ratway, and A. F. Baker, "Lurgi Gasifier Tests of Pennsylvania Anthracite," U.S. Dept. of Interior, U.S.B.M. Rept. of Inv. 5240, 1958.

13 G. M. Hamilton, "Gasification of Solid Fuels in the Wellman-Galusha Gas Producer," presented at the annual meeting of AIME, St. Louis, MO, March 1961.

14   Anon., A Current Appraisal of Underground Coal Gasification,
     Cambridge, MA: Arthur D. Little Inc. (NTIS - PB - 209 - 274)
     1972.

15   G. H. Lamb, Underground Coal Gasification, Noyes Data Corp.,
     Park Ridge, NY, 255 p., 1977.

16   S. B. King, "Composition of Selected Fractions from Coal Tars
     Produced from an Underground Coal Gasification Test," Amer.
     Chem. Soc. - Fuel Chemistry Division Preprints, Vol. 22,
     No. 2, March 1977.

17   Anon., "U.S. Bureau of Mines Study of Low and Intermediate
     Btu Gas from Coal for Iron Ore Pelletizing," McKee, p. 2,
     1977.

18   S. W. Herman, et al., Energy Futures, Cambridge, MA:
     Ballinger Publ. Co., p. 401, 1977.

19   Anon., "Coal Conversion Activities Picking Up," Chemical
     and Engineering News, p. 25, Dec. 1, 1975.

20   B. Lewis and G. Von Elbe, Combustion Flames and Explosions
     of Gases, (2nd ed.), New York: Academic Press, pp. 705-706,
     1961.

CHAPTER 5

MEDIUM BTU GAS FROM COAL

5.1 INTRODUCTION

It was observed, in the case of producing low Btu gas, that an efficiency-environmental trade-off had been made. Gasification consumes more coal, water, and other resources in the production of a clean industrial fuel. An extension of that trade-off exists in the case of medium Btu gas ($\sim$300 Btu/ft^3) production, however, the trade-off is between low and medium Btu gas. If the latter is to be chosen, some efficiency is sacrificed in order to obtain the following operational advantages.

1. The ability to produce an inherently cleaner, essentially nitrogen-free gas.
2. The ability to use the gas with less deleterious effect on boiler rating.
3. The ability to burn the gas after it has cooled down with less effect on its total energy content.
4. The ability to store the gas, in modest quantities and for reasonably short periods of time, to smooth out the various peaks and valleys of gasification and product manufacturing operations.
5. The ability to transport the fuel over appreciable distances (e.g., 15 to 25 miles) economically.

In addition to the efficiency penalty there are economic costs associated with producing medium Btu gas. The minimum economic size of operations is significantly larger, requiring the raising of more capital. The construction of an oxygen producing plant is yet another capital investment that must be borne if such a gas is to be manufactured.

Despite the lower efficiencies and higher capital costs, the advantages of medium Btu gas may make it attractive, particularly in retrofit situations at large installations. Production systems exist, now, for producing this gas on a commercial basis. These include the Lurgi, Koppers-Totzek, and Winkler units plus second generation systems designed to feed substitute natural gas complexes. The gas quality from such units has been defined precisely, and retrofit procedures have been well established.

Because medium Btu gas now exists as a viable alternative clean industrial fuel, it merits careful examination here.

5.2   PRODUCTION SYSTEMS FOR MEDIUM BTU GAS

Conversion of coal to medium Btu fuel gas consists of essentially the same reactions and gasifiers as those for low Btu gas. The principal products are carbon monoxide and hydrogen. The additional requirement for production of low nitrogen fuel gas is either oxygen enriched air or oxygen instead of air as a reactant to be combined with steam and coal. The major improvements since the 1940s that make medium Btu gas more promising include improved materials of construction for gasifiers, new methods of solids-gas contacting (fluidized beds, entrained flow and ebulating beds), and the production of oxygen on a commercial scale. Since gas delivery pressure can be an important consideration, especially when the gas is used in a turbine, operation of gasification systems at elevated pressures has also developed.

5.2.1   Production Approaches Available

Table 5.1 lists the gasifier types and characteristics for production of medium Btu gas [1]. As indicated nonfuel gases, including N_2, H_2S, COS, and Ar, usually amount to less than 3 percent when oxygen and steam are fed to the gasifiers. The information given in Table 5.2 is of perhaps more value in assessing the status of the proven technology for medium Btu gasification. It is likely that most of the commercially proven processes will be utilized for several years unless a clear advantage for one or more becomes evident.

Figures 5.1, 5.2, and 5.3 give the essential features of the gasification systems from Table 5.2. Entrained flow, fluidized bed, and fixed bed operation are represented in these commercial systems. The important gasification reactions are similar to those given in Chapter 4 for low Btu gasification. However, the absence of a large volume of nitrogen in the feed

TABLE 5.1 Medium Btu Gasifiers and Characteristics

Gasifier Type and Licensor	Feed Requirements	Operating Conditions		HHV (Btu/ft^3)
		T(°F)[a]	P(psig)	
Fixed (Moving Bed)				
Lurgi (American Lurgi Corporation	1/8" × 1-1/2" act[b]	1200–1500	350–450	285
Woodall-Duckham/ Gas Integrale	1/4"–1-1/2" (coals w FSI>2-1/2)	2200	atm	280
Entrained Flow				
Babcock & Wilcox	-200 mesh (79–90%)	3400	atm to 300	300
Bigas (Bituminous Coal Research Inc.)	-200 mesh (70%)	2700–3000	500–1500	356
Koppers-Totzek (Koppers Co., Inc.)	-200 mesh (70–90%)	3500	atm	286
Texaco (Texaco Development Corporation	pulverized	>ash fusion pt.	300–1200	253

Fluidized Bed

Battelle/Carbide (Battelle Memorial Institute)	crushed	1800	100	
CO_2 Acceptor (Conoco Coal Development Co.)	-8 + 100 mesh	1500	150	380
COGAS (COGAS Development Co.)	0 × 1/8"	1600	15-45	335
HYGAS (Institute of Gas Technology)	-10 mesh	1850	~1160	375
Synthane (DOE-Pittsburgh Energy Technology Center)	-20 mesh	1800	1000	355
Winkler (Davy Powergas Inc.)	0 × 3/8"	1800	atm	388

Molten Media

AL Molten Salt Atomics International	-1/4"	~1800	up to 280	300
Otto Rummel C. Otto & Co. G.M.B.H.	-16 mesh	~1700	360	278

(Continued)

TABLE 5.1 (Continued)

| Gasifier Type and Licensor | Gas Composition (Dry%) | | | | | |
	CO	CO_2	H_2	Nonfuel gases	CH_4	Other HC
Fixed (Moving) Bed						
Lurgi (American Lurgi Corporation)	16.9	31.5	39.4	2.4	9.0	0.8
Woodall-Duckham/Gas Integrale	37.5	18	38.4	2.6	3.5	–
Entrained Flow						
Babcock & Wilcox	65.3	5	27.9	1.8	–	–
Bigas (Bituminous Coal Research, Inc.)	29.3	21.5	32	1.5	15.7	–
Koppers-Totzek (Koppers Co., Inc.)	52.2	10	36	1.5	–	–
Texaco (Texaco Development Corporation)	37.6	20.8	39	2.1	0.5	–
Fluidized Bed						
Battelle/Carbide(Battelle Memorial Institute)						
CO_2 Acceptor (Conoco Coal Development Co.)	15.5	9.1	58.8	2.9	13.7	–

COGAS (CoGas Development Co.)	31.2	6.6	57.9	0.3	4.0	–
HYGAS (Institute of Gas Technology)	26.1	24.1	30.7	0.8	16.6	1.3
Synthane (DOE-Pittsburgh Energy Research Center)	13.2	36.2	32.3	1.6	15.0	1.6
Winkler (Davy Powergas Inc.)	48.2	13.8	35.3	0.9	1.8	–
Molten Media						
AL Molten Salt (Atomics International)	–	–	–	–	–	–
Otto Rummel C. Otto & Co. G.M.B.H.	53.6	14	30.7	1.2	0.5	–

[a] To convert to °C use

$$°C = \frac{5°C}{9°F} \ (°F = 40°F) - 40°C$$

[b] act = all coal types can be treated.

75

TABLE 5.2 Status of Commercial Process for Medium Btu Gas

| | | Commercial Status | |
| | | | |
Process	First Commercial Plant(s)	Number of Plant(s) Built or Ordered (to 1977)[a]	Number of Gasifiers[a] (1977)
Koppers-Totzek Koppers Company, Inc. Pittsburgh, PA	Finland 1952	20 (2)	49
Lurgi Lurgi Gesellschaften Frankfurt (am Main) West Germany	Herschfelde, Germany 1936	18 (3)	69
Woodall-Duckham/Gas (Woodall-Duckham USA Limited, Pittsburgh, PA)	– ∿1927 (cyclic)	–	100 (air blown)
Integrale Il Gas Integrale Milan, Italy	∿1947	–	15 (O_2-blown)
Texaco Texaco Development Corp., New York, New York	1953	70 (2)	–
Winkler Davy Powergas Inc. Lakeland, Florida	Leuna, Germany 1926	16 (4)	36

[a]The number of plants and gasifiers are total numbers not restricted to those using coal as a fuel or to medium Btu gas as a product.

76

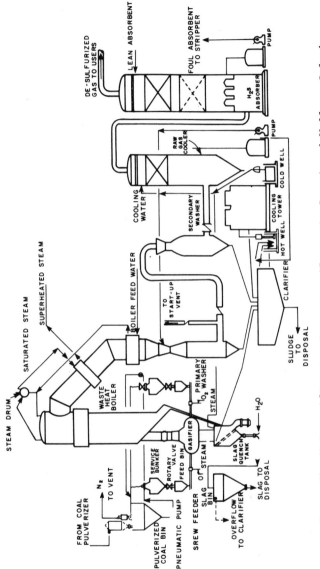

Figure 5.1. The K-T Gasification Process. Where the Lurgi and Wellman-Galusha reactors are countercurrent systems, the Koppers-Totzek process employs an entrained flow reactor. It can handle all kinds of coal. Size of 250-450 tons/day of coal are common. The pure oxygen size ensures the production of medium Btu gas rather than low Btu gas. Presently 20 such units are operated around the world.

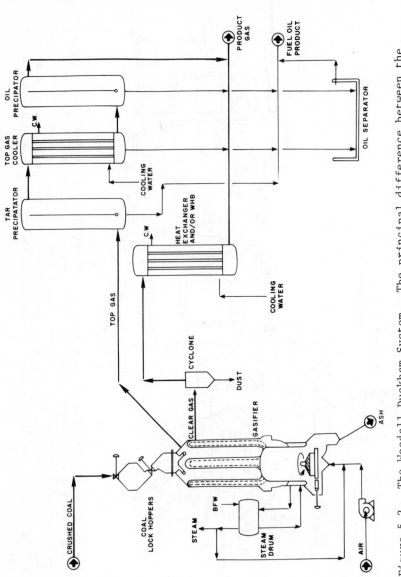

Figure 5.2. The Woodall Duckham System. The principal difference between the Woodall Duckham and previous systems described is the fact that it is air blown and produces both oil and synthesis gas products.

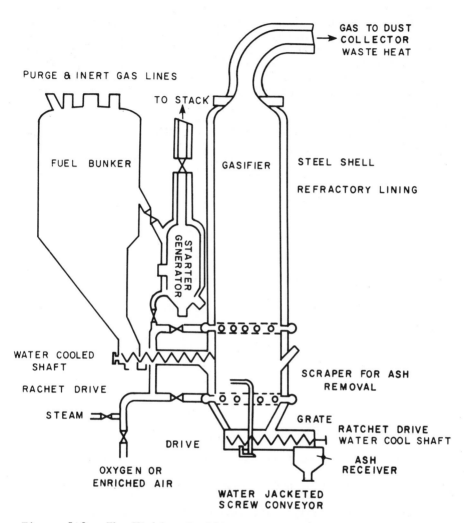

PURGE & INERT GAS LINES

FUEL BUNKER

TO STACK

GASIFIER

GAS TO DUST
COLLECTOR
WASTE HEAT

STEEL SHELL

REFRACTORY LINING

STARTER
GENERATOR

WATER COOLED
SHAFT

RACHET DRIVE

SCRAPER FOR ASH
REMOVAL

STEAM

GRATE

RATCHET DRIVE
WATER COOL SHAFT

DRIVE

ASH
RECEIVER

OXYGEN OR
ENRICHED AIR

WATER JACKETED
SCREW CONVEYOR

Figure 5.3. The Winkler Gasification Process. The Winkler is
the only fluidized bed gasifier commercially available. The
fluidized bed is the distinguishing characteristic of this
system, first built in 1926.

gas changes the heat balance requirements. Continuous operation is easier than air blown gasification since the heat for gasification reactions can be supplied in the gasifier by oxidation of part of the coal. This continuous operation and decreased heat loss increases the gasification efficiency making possible increased rates of gas production. Operation at pressures above atmospheric is also more attractive using oxygen.

The basic steps in gasification to medium Btu gas are the same as those for producing low Btu gas; namely, drying and heating, devolatilization, and chemical reaction.

The chemical reactions important in the production of CO and hydrogen from steam, oxygen and coal include the following [5]:

$$C + 1/2\ O_2 \quad \rightarrow \quad CO \tag{1}$$

$$C + O_2 \quad \rightarrow \quad CO_2 \tag{2}$$

$$C + CO_2 \quad \rightarrow \quad 2CO \tag{3}$$

$$CO + H_2O \quad \rightarrow \quad CO_2 + H_2 \ \text{(water gas shift)} \tag{4}$$

$$C + H_2O \quad \rightarrow \quad CO + H_2 \tag{5}$$

Some gasifier/coal combinations result in significant amounts of methane in the product gas (Hygas, Bigas, Synthane and CO_2-Accepter processes). The following reactions are involved in the formation of methane:

$$C + 2H_2 \quad \rightarrow \quad CH_4 \tag{6}$$

$$CO + 3H_2 \quad \rightarrow \quad CH_4 + H_2O \tag{7}$$

$$CO_2 + 4H_2 \quad \rightarrow \quad CH_4 + 2H_2O \tag{8}$$

Some methane and other hydrocarbons are undoubtedly formed as a result of pyrolysis of volatile matter from coal. These are usually present in only small amounts without catalysts or other conditions designed for methanation. The temperatures used in medium Btu gasification vary from less than 1500°F to 3500°F. At these temperatures the stable products are predominantly CO and H_2 and the reactions leading to these components are largely endothermic. The amount of heat required is so large that every effort must be made to utilize the waste heat in steam boilers to provide gasifier process steam.

5.2.2 Commercial Systems Available

Although complete gasification of the solid coal is desirable this is not usually achieved to more than 85 to 90 percent. Char left after gasification can be burned as a supplemental fuel for steam boilers. Specific characteristics of the commercial processes for medium Btu gasification are discussed since these represent different types of gasifiers and considerable data and experiences are available. This discussion is limited to the three processes that have been used commercially with coal as a feedstock; namely Winkler, Lurgi, and Koppers-Totzek. It must be recognized that product compositions and other data for each process are representative only and depend upon the actual process conditions and the feed materials.

1. Winkler Gasifiers. The Winkler is a fluidized bed unit that was the first commercial large scale process used to gasify lower rank coals and char. A significant advance of the Winkler was continuous operations. Earlier gas producers typically operated with blow and run cycles with fuel gas produced only during the run cycle.

The earliest Winkler units installed at Leuna, Germany in 1926 were air blown, but after being idle for about two years these units were restored to production using steam and oxygen. Since 1933 all commercial Winklers have been blown with oxygen and steam, producing medium Btu fuel gases.

    a. Advantages of Winkler units.

        i. Operation of part load possible.
        ii. Ease and speed of startup.
       iii. Gas product low in tars.
       iv. Pulverizing and specific size screening
           is unnecessary.
        v. Good temperature control.
       vi. Low concentration of contaminants in product
           gas (i.e., HCN, ammonia).

    b. Disadvantages of Winklers.

        i. Difficulty of operation with bituminous and
          anthracite coals.
        ii. Difficulty of direct feed of caking coals.
       iii. Poor kinetics.
       iv. Limited temperature range for operation ($\sim$1900°F).
       v. High dust carry over (including carbon and ash).

2. **Lurgi Gasifiers.** Lurgi gasifiers as shown previously in Chapter 4, first commercially operated in 1936, contributed several features that have been valuable in the development of coal gasification [1], [3], [6]. The most important of these was gasification at a pressure that allowed for higher rates of production, the use of smaller coal sizes and a higher heating value in the product gas (due to higher methane content). Having the fuel gas at process pressure can also facilitate removal of CO_2 and H_2S, thereby further increasing the quality of the product gas. Since the Lurgi gasifiers first developed were operated dry (with temperatures below the ash fusion temperature to avoid clinkering) the gasifier size limited the coal processing capacity. The largest commercial Lurgi units could, in fact, process less than 600 tons of coal per day. The dry ash Lurgi also has the following additional disadvantages:

    a. High steam/oxygen ratios are necessary to control the temperature below the clinkering temperature. This results in much of the feed steam appearing in the products of the reaction that decreases the gasifier efficiency.

    b. The product gases contain considerable amounts of tars.

    c. To avoid clinkering, even when the average temperature is below the ash fusion point, a mechanical distributor is used. This along with other moving parts in the gasifier make for a complex system subject to many operational problems.

    d. Lurgi gasifiers experience operational difficulties with some fuels, particularly caking coals.

Some of the foregoing problems can be avoided by operation of the Lurgi process at higher temperatures; that is, under slagging conditions. Since the early 1950s Lurgi Gesselschaft and others have performed tests to determine the validity of such techniques. The following advantages were determined for slagging Lurgi units over dry Lurgi's:

    a. Higher thermal efficiency.

    b. Lower steam requirements (less by a factor of approximately 5).

    c. Better adaptability to fuels of different characteristics.

    d. Decreased liquor production (due to less H_2O and tars in the products).

    e. Higher coal processing capacity as a result of the higher temperatures employed and the decreased flow velocity of the feed gases.

As with all changes in process conditions while some advantages are gained others are lost. This is true for the slagging Lurgi gasifier, which has the following drawbacks.

    a. More demand on construction materials of the gasifier and slag handling components of the system due to the higher temperatures.

    b. Difficulties with processing unconsolidated or highly friable coals.

    c. There is a strong likelihood that highly caking coals could not be gasified without plugging problems even with mechanical stirring of the bed.

Despite some problems with slagging Lurgi's they have great potential for gasification of a wide variety of fuels and interest is high in further development. This is demonstrated by the current support of slagging gasifier development at West-field, England by 15 American companies and organizations, as well as the British Gas Council. Early results look promising with coal processing rates over five times as high for a slag-ging reactor (1210 lbs of coal/ft^2-hr versus 226 lbs of coal/ft^2-hr for in a dry ash Lurgi unit).

3. **Koppers-Totzek Gasifier Units.** The Koppers-Totzek (K-T) process is a relative newcomer among those commercially in operation. The first pilot plant was built in Louisiana, Missouri as a result of joint efforts of Koppers-Essen and Koppers-Pittsburgh under arrangements with the U.S. Bureau of Mines. This pilot plant was operated for approximately two years beginning in May of 1949. It was used mainly to supply synthesis gas to a pilot plant operated for the production of liquid hydrocarbons (Fischer-Tropsch) [2].

The first commercial gasifier units (3 gasifiers) were built in Oulu, Finland in 1952. Since then plants in Japan, Spain, Belgium, Portugal, Greece, the United Arab Republic, Thailand, Turkey, East Germany, and Zambia have been completed. Others are under construction.

The K-T construction is an entrained-flow gasifier operating at atmospheric pressure using oxygen, steam, and pulverized coal (70 percent through 200 mesh). The fine coal is fed through mixing nozzles with oxygen at high velocity (300 feet/sec). This high rate prevents flame propagation back into the feed nozzles. Coal processing capacity varies depending on the number of burner heads used. A four headed gasifier unit with burners at 90 degrees from each other can process about 850 tons of coal per day. The coal ash is about equally distributed between the solids at the bottom of the gasifier and the product gases.

The refuse must be quenched by water and the product gases by water sprays when they contain ash. Carbon conversions have been reported over 99 percent for lignites, but lower for higher rank coals and chars (~90 percent). There are several advantages of K-T units and the commercialization of the process during a period of highly available and cheap fuels attest to this. The following advantages should be noted.

    a. Good safety record (no major explosions).
    b. The K-T gasifier is relatively simple and flexible.
    c. Little effect on the gasification process due to
        physical properties of the feed coal, such as fines,
        ash fusion temperature, ash content, etc.
    d. Rapid start-up and shut-down capability.

Disadvantages of the K-T process include:

    a. Low thermal efficiency.
    b. Limitation of feed coal to <10 percent moisture.
    c. Lack of data on high rank coals.
    d. Relatively high cost of energy.
    e. Operation with air and steam is not promising.

5.3 USE CHARACTERISTICS OF MEDIUM BTU GAS

The major components of medium Btu gas are CO, H_2, and CO_2. Since CO and H_2 are the same fuel components contained in low Btu fuel gas one could expect that the same applications would apply. This is partially true so that use of medium Btu gas would be applicable to electric power generation and to process heat or steam. Additionally medium Btu gas can be used as a synthesis gas for the production of methane, higher hydrocarbons, or other products by Fischer-Tropsch or related synthesis.

The major users of natural gas, coke oven gas, refinery gas, and residual fuel oil that could utilize medium Btu gas in process heaters or boilers are petroleum refiners and blast furnace and steel companies [7].

5.3.1 Retrofit Characterists and Problems

When medium Btu gas is substituted for natural gas some problems in the burners may occur. Because of the lower heating value of medium Btu gas compared with natural gas (900 to 1000 Btu/SCF) or coke oven gas (~500 Btu/SCF) flow distortions in nozzle mix burners can lead to unsatisfactory performance or

flashback. New burner designs are now available to correct this
problem, but others may be required for some applications. In
some cases currently used burners (i.e., process heater burners)
that have high turndown ratio can use a wide variety of fuel
gases. These might be adjusted to use medium Btu gas. In a
refinery conversion from natural gas to medium Btu gas may
require replacement of all burners. Replacement of boilers may
also be more economical than retrofitting of the natural gas
boilers.

While a complete discussion of the fuel stability factors
and combustion characteristics is beyond the scope of this chapter some conclusions are offered. The reader is referred to a
more complete treatment that includes data on fuel composition
and thermal properties, fuel stability factors, flashback velocity gradients, and burner design types and characteristics [8].

The important factors in retrofitting for combustion of a
medium (or low) Btu gas in place of natural gas or oil include
the following: 1) burner type and design, and 2) fuel and air
supply lines to burners. The following list shows three general
types of burners used for gas combustion.

1. Premix Burners that are supplied with fuel gas and air
    that have been mixed before the mixture enters the
    burner nozzle.
2. Delayed-Mixing Burners, where the air and fuel gas leave
    the burner nozzle unmixed. The result is a long diffu-
    sion flame since most of the air-fuel mixing is by
    diffusion.
3. Nozzle-Mix Burners, where the air and fuel gas is mixed
    just as they leave the burner port. Mixing is rapid so
    that nozzle mix burners burn more intensely than
    delayed mixing burners.

As indicated by these burner types flames are either pre-
mixed or diffusion types. The flame stability is a function of
the flame characteristics. For example, a flame may be laminar
or turbulent, or contain both types of mixing and combustion
characteristics.

The changes that result from using a fuel gas different in
composition and heating value from another fuel are primarily
in the temperature attained by the flame (and thus the radiation
and convection of heat to furnace surfaces).

Other changes necessary for utilization of medium (or low)
Btu fuel gas in place of natural gas are related to the flow
requirements of the burner systems. If the same fuel lines are
used the flow rate for medium Btu gas must be three to four

times greater than that for natural gas (in order to supply the
same heat per unit time). The increased flow rates are accom-
panied by large pressure drops which can cause serious problems
in retrofitting.

The solution to the flow problems would be one of the
following.

1. To allow for the large pressure drop, fuel gases could
   be supplied under pressure. (Presently only Lurgi
   gasifiers could supply such gases.) This solution has
   the serious drawback of having carbon monoxide, a
   highly toxic gas, under pressure. Any leakage of the
   fuel gas would pose a hazard to workers.
2. The fuel supply lines to burners could be enlarged to
   give increased quantities of fuel gas at reduced flow
   velocity. This could be an expensive solution where
   the lines are extensive or impractical where limited
   space is available.
3. The amount of heat being supplied to a boiler or heater
   could be reduced. This derating has been estimated as
   up to 5 percent for medium Btu gas with a heating value
   of 300 Btu/SCF and up to 25 percent for low Btu gas with
   a heating value of 150 Btu/SCF [8].

This particular choice among the above solutions to the
flow problems of retrofitting and installation using natural gas,
or other fuel, will depend on the particulars of the installation. Table 5.3 compares the flow rates and pressure drops for
some manufactured gases with natural gas.

5.3.2 Environmental Aspects of Medium Btu Gas Use

Any substitution for natural gas or oil by medium Btu gas
produced from coal would involve the usual emission problems of
coal utilization, namely the following:

Dust and leachate from coal storage piles

Dust from coal handling, drying and pulverizing
operations.

In addition the gasification of coal to medium Btu gas with
attendant gas cleanup and combustion would involve the following
pollutants:

TABLE 5.3. Relative Flow Rates and Pressure Drops for Manufactured and Natural Gas[a]

Fuel Gas	Higher Heating Value (Btu/scf)		Flow Rate			Pressure Drop		
	Fuel Gas	Gas + Air (stoich. mixture)	Fuel	Mixture	Products	Fuel	Mixture	Products
Lurgi (oxygen blown)	322	87	3.31	1.11	1.03	13.5	1.19	1.11
Koppers-Totzek (oxygen blown)	294	93	3.63	1.04	0.885	14.9	1.01	0.806
Wellman-Galusha (air blown)	172	94	6.20	1.31	1.18	52.7	1.66	1.49
Natural Gas (methane)	1066	97	1.0	1.0	1.0	1.0	1.0	1.0

[a]Adapted from [8, Appendix A, p. A28]. Relative flow velocities and pressure drops are given as a factor of that for natural gas (using unity for natural gas) with turbulant flow assumed.

	Discharge Pollutants	Amounts
Oxygen plant	N_2, CO_2, H_2O, O_2	trace, except N_2
Cooling tower	H_2O, suspended and dissolved solids	trace amounts of solids
Claus process for sulfur		
tail gas	CO_2, H_2S	up to 3 percent H_2S in tail gas
solids	sulfur	
Scrubbers	tar, NH_3, phenols, HCN, hydrocarbons	–
Combustion	SO_2, COS, CS_2, CO, NO_x, particulates, organics	–

It should be emphasized that the total "emissions" from a coal combustion operation and a coal gasification and gas combustion plant involve the same elements in approximately the same amounts. However, those emissions must be controlled because of the possibility of atmospheric or effluent pollution is significantly changed by the gasification step. For example, in order to meet the federal standard for solid fuel firing (1.2 lbs of $SO_2/10^6$ Btu) a bituminous or subbituminous coal would have to contain less than 0.5 to 0.8 percent sulfur (without desulfurization). To meet the same standard a fuel gas with 300 ppm sulfur concentration could have a heating value less than 100 Btu/SCF. Approximately the same amount of sulfur is involved, but in the coal gasification significant quantities of sulfur are produced as a solid byproduct and are thus not present in the fuel gas.

The particular method of fuel gas utilization would determine the environmental impact of a plant. As indicated earlier in Chapter 4 considerable reduction in emissions to the atmosphere and a reduction in cooling tower heat rejection is possible by utilization of gaseous fuels from coal. In the case of medium Btu gas production and a combined cycle power plant (1000 MW) relative to a pulverized coal boiler with stack gas scrubbing the reduction in particulates would be insignificant; however, there would be a 90 percent reduction in NO_x, a 65 percent reduction in cooling tower heat rejection, and only about 10 percent of the land for disposal of wastes would be required.

REFERENCES

1   U.S. Dept. of Commerce (National Technical Information Service) prepared by Dravo Corporation, "Handbook of Gasifiers and Gas Treatment Systems," prepared for the U.S. Energy Research and Development Administration, Feb. 1976.

2   J. F. Farnsworth, H. F. Leonard, D. M. Mitsak, and R. Wintrell, "K-T-Koppers Commercially Proven Coal and Multiple-Fuel Gasifier," presented to the Association of Iron and Steel Engineers 1974 Annual Convention, Philadelphia, PA, April 1974.

3   P. F. H. Rudolph, Oil Gas J, Vol. 71, No. 4, pp. 90-92, 1973.

4   M. Goodman, E. Bailey, "Synthetic Medium Btu Gas via Winkler Process," from 4th Annual International Conference on Coal Gasification, Liquefaction and Conversion to Electricity, University of Pittsburgh, August 1977.

5   H. H. Lowry (Ed.), Chemistry of Coal Utilization, Supplementary Volume, NAS-NRC. New York: John Wiley and Sons, 893-904, 1963.

6   J. Yerushalmi, "Report on EPRI's Workshop on Clean Gaseous Fuels from Coal," prepared for the Electric Power Research Institute, January 1977.

7   R. A. Ashworth, K. C. Vyas, and D. G. Bonamer, "Study of Low and Intermediate Btu Gas from Coal for Iron Ore Pelletizing," Arthur G. McKee & Company, Cleveland, Ohio, prepared for U.S. Dept. of Interior, Bureau of Mines, 181 p., March 1977.

8   B. A. Ball, A. A. Putnam, D. W. Hissong, J. Varga, B. C. Hsieh, J. H. Payer, and R. E. Barrent, "Environmental Aspects of Retrofitting Two Industries to Low- and Intermediate-Energy Gas from Coal," Battelle Columbus Labs, U.S. Dept. Commerce, National Technical Information Service, PB 253-946, Appendix A, A1-A30, April 1976.

CHAPTER 6

HIGH BTU GAS FROM COAL

6.1 INTRODUCTION

While the production of low and medium Btu gas from coal provides a mechanism for responding to the needs of industry, the production of high Btu (900 Btu+ft^3) gas answers the more ubiquitous problem of how to respond to natural gas shortages. That fundamental difference creates a series of technical and physical distinctions between high Btu gas and the less concentrated coal based gaseous fuels. These distinctions can be summarized as follows:

1. Location of production unit.
2. Minimum economic size of production unit.
3. Complexity of process train.
4. Fuel use characteristics.

Specific economic and environmental issues associated with the production of SNG will be dealt with in Chapter 9.

It should be emphasized here that the production of synthesis gas and/or medium – Btu gas may be regarded as a first step in the manufacture of substitute natural gas (SNG) or high Btu gas. Essentially any gasifier or process that produces medium Btu gas can also produce SNG when appropriate methanation is implemented. Gas product properties that are important in the production of SNG are 1) the ratio of H_2 to CO in the gas produced, and 2) the amount of methane already contained in the product from the gasification. All of the various medium Btu gasification processes discussed in Chapter 5 could be used to produce SNG. However, the two gas product properties referred to above determine the

real relative applicability of a process to commercial production
of SNG by the addition of shift conversion and methanation steps.
     The utility structure employed by natural gas industry has
created the need for this substitute natural gas (SNG) and has
caused the size and location distinctions as well. The conversion of homes and industries from locally produced "town gas" to
natural gas resulted in this country in the construction of a system including over 260,000 miles of gas transmission pipeline
plus some 650,000 miles of gas distribution pipelines serving 41
million residential and 3.5 million commercial and industrial
customers. This system represents a multi-billion dollar investment that, as Chapter 1 points out, is facing serious underutili-
zation. This massive system also makes a return to the decentral-
ized town gas system virtually impossible.
     To meet the needs of nearly 45 million customers, natural
gas companies are designing SNG plants. Already 17 peaking plants
that convert the light fractions of refined petroleum into SNG
exist. The more fundamental solution is base load SNG plants
designed to convert coal.
     Like coal liquefaction plants, these SNG plants are essen-
tially large refinery complexes located at the point of coal production rather than the point of coal consumption. Their output
is targeted for the national energy distribution system rather
than a specific plant. Thus, for example, the El Paso Gas pro-
posal was designed for Farmington, New Mexico. There the basic
production needs—coal, water, etc.—could be met.
     In addition to the location distinction, size is an important
differentiation between industrial and SNG gas production. In
the case of the former, Wellman-Galusha units operating on 45
tons/day (tpd) are employed; Koppers-Totzek units of 450 tpd are
not uncommon. Such units produce gas containing 1×10^9 to
10×10^9 Btu/day. This is in stark contrast to the standard high
Btu gas size of $+250 \times 10^6$ ft^3 plant producing $+250 \times 10^9$ Btu/day.
In the latter case, only the large plants can have a significant
impact on the distribution system. Given those institutional differences, it is now necessary to examine the technological issues
in more detail.

6.2 PROCESS TECHNOLOGIES

     The process for refining coal into SNG begins, in all cases,
with medium Btu gas production as described in Chapter 5.
Because the specific composition of this gas is critical to the
performance of SNG production units, compositions from various
gasifiers are reproduced here as Table 6.1. As is shown in Table
6.1 production of nonfuel gases (principally H_2S) ranges from

TABLE 6.1 Representative Synthesis Gas Compositions for Producing SNG[a]

Gasifier	Gas Composition (dry%)					
	CO	CO_2	H_2	CH_4	C_xH_x	Other Nonfuel Gases
Lurgi	16.9	31.5	39.4	9.0	0.8	2.4
BiGas	29.3	21.5	32.0	15.7	0.0	1.5
Koppers-Totzek	52.2	10.0	36.0	0.0	0.0	1.5
Texaco	37.6	20.8	39.0	0.5	0.0	2.1
CO_2 Acceptor	15.5	9.1	58.8	13.7	0.0	2.9
HyGas	26.1	24.1	30.7	16.6	1.3	0.8
Synthane	13.2	36.2	32.3	15.0	1.6	1.6
Winkler	48.8	13.8	35.3	1.8	0.0	0.9

[a]Adapted from Chapter 5.

0.8 percent (dry percent) to 2.9 percent. Production of CH_4
ranges from 0.0 percent to 16.6 percent.

In the United States by the mid 1970s over 20 conversion
processes were in various stages of research and development,
including the European processes. European technology was further
along because of the higher level of interest in gasification,
since natural gas was less available than in the United States.
Some of the processes that have been specifically supported by
the U.S. Government and/or industry for the production of SNG
included HYGAS (Institute of Gas Technology), Bi-Gas (Bituminous
Coal Research Inc.), CO_2-Acceptor Process (Conoco Coal Development Co.), Synthane (U.S. Bureau of Mines, later the Department
of Energy), COGAS (FMC Corporation, later COGAS Development Company and subsequently by the Illinois Coal Gasification Group),
Exxon Catalytic Coal Gasification Process (Exxon Research and
Engineering Company), Hydrane Process (U.S. Bureau of Mines, later
the Department of Energy), and some of the European technologies
adapted to U.S. coals (such as the Slagging Lurgi Process developed by Conoco). Additionally, as shown in Table 6.1, any process
that produces synthesis gas that could be methanated is a candidate for production of SNG. The variety of process include those
that produce little methane in the gasification step (Koppers-
Totzek, Texaco, and Winkler), those that use direct hydrogasifica-
tion (Hydrane), those that produce considerable methane in the
gasification step, but also require methanation (Lurgi, Bi-Gas,
CO_2-Acceptor, Hygas, Synthane), and one that uses a catalyst in
the gasification step (Exxon).*

6.2.1 Basic Process Description

Methanation reactions occur to some extent in the formation
of synthesis gas. In fact, one might ask in the case of producing high Btu gas from coal, "Why not produce methane directly
from coal?" This would be an ideal situation if it could be
done:

$$Coal + H_2O \rightarrow CH_4 + CO_2 \tag{1}$$

This reaction (as pointed out by G. A. Mills) [1, 2] is
accomplished by bacteria, although cellulose is substituted for
coal. The material and heat balance of (1) is satisfactory but
the reaction cannot be effected at a reasonable rate at ambient

---

*W. R. Epperly and H. M. Siegel, "Catalytic Coal Gasification for
SNG Production," in Proc. of 11th Intersociety Energy Conversion
Eng. Conf., Stateline, Nevada, Sept. 1976.

conditions. As the temperature is raised to high values, where gasification reactions proceed at adequate rates, CH_4 is not a stable product. In other words, the conditions for favorable reaction between coal and steam are not the favorable conditions for methane formation. Thus, the technology for coal gasification to high Btu gas has been developed involving three steps: (1) Gasification;

$$\text{Coal + steam} \quad \xrightarrow{\text{(endothermic)}} \quad CO_2 + H_2/CO \text{ and some } CH_4 \qquad (2)$$

(2) synthesis gas shift (to H_2/CO ratios that are more appropriate for efficient methane production; usually about 3:1);

$$CO + steam \rightarrow 3H_2 + CO \qquad (3)$$

and (3) methanation;

$$3H_2 + CO \quad \xrightarrow{\text{(exothermic)}} \quad CH_4 + H_2O \qquad (4)$$

Other methanation reactions include:

$$CO_2 + 4H_2 \rightarrow CH_4 + 2H_2O \qquad (5)$$

and

$$C + 2H_2 \rightarrow CH_4 \qquad (6)$$

Several other steps are usually necessary in the gasification of coal to SNG. They may include pretreatment (in the case of caking coals) and removal of "acid gases" (CO_2, H_2S, etc.). There are also other reactions possible at gasification conditions that take place to a greater or lesser extent depending on the exact gasification scheme used; such as:

$$C + CO_2 \rightarrow 2CO \qquad (7)$$

$$\tfrac{1}{2}O_2 + H_2 \rightarrow H_2O \qquad (8)$$

$$CO + 2H_2 \rightarrow CH_3OH \text{ (methanol)} \qquad (9)$$

$$CO_2 + 4H_2 \rightarrow CH_4 + 2H_2O \qquad (10)$$

$$C + 2H_2O \rightarrow 2H_2 + CO_2 \qquad (11)$$

$$2C + H_2 \quad \rightarrow \quad C_2H_2 \qquad\qquad (12)$$

$$CH_4 + 2H_2O \rightarrow \quad CO_2 + 4H_2 \qquad\qquad (13)$$

The important methanation reaction is best carried out using a catalyst. For example, reaction (4) was catalyzed very early (1902 Sabatier and Senderene) [4] by nickel. Other catalysts for this reaction include: iron, cobalt, ruthenium, rhenium, palladium, ozmium, indium, platinum, and silver. Recently research results have been obtained that show that alkaline catalysts (sodium and potassium oxide) have been tested and shown to greatly accelerate the gasification reactions [1]. Possibly new approaches to high Btu gas production from coal could result from such catalysis. A treatment of the development of methanation catalysts for SNG processes is available to outline poisoning and other important factors [5]. To accomplish the necessary reactions for high Btu gas (methane production) the following take place in the gasifier: elimination of caking (if caking coal is used), coal devolatilization (some methane is produced in this step), reaction of devolatilized coal with hydrogen-carbon monoxide and steam (some additional methane formed), and shifting of the synthesis gas produced to increase the synthesis of CH_4 from CO and H_2.

The various gasification processes are only different in approach to accomplish these steps. The classification of such "SNG processes" can be made by dividing into the following:

1. Systems that carry out the heat abosrbing reaction (3) to give CO and H_2 <u>and</u> the reaction of devolatilized coal in one unit together. Heat for the reaction can be supplied by electrical heating, supplying hot steam, by partial oxidation of the carbon in the coal by oxygen,

$$(i.e., \ C + 1/2 \ O_2 \rightarrow CO \qquad - 26.6 \ \frac{Kcal}{mole} \ ) \qquad (14)$$

   or by heat transfer by hot fluid or solids to the gasifier bed.

2. Systems that carry out 1, but that supply heat for these endothermic reactions by addition of oxygen to feed steam (this, however, renders the $(H_2 + CO)/(H_2O + CO_2)$ ratio less favorable than that in 1).

3. Systems that add hydrogen to the steam before the steam-char reaction; this results in higher hydrogen/CO ratios as well as $(H_2 + CO)/H_2O + CO_2)$.

The gasifier types and conditions are mostly those previously described in Chapters 3 and 4 for low-Btu and medium-Btu gasification and shown in Table 6.1. The major distinction being the addition of a catalytic methanation reactor. Commercial operation alone can really answer the questions about the "best method" in terms of economics. However, it appears that gasifiers that produce the most methane by noncatalytic reactions in the gasifier have an advantage since less methane is then required from the higher cost catalytic methanation step.

Gasification and methanation temperature requirements are important and different as illustrated by the following data [5].

Reaction	$- H$ at 927°C (Kcal/mole)	Temperature when Kp 1(°C)	(°F)
(a) $C + H_2O_{(v)} \rightarrow CO + H_2$	-32.46 (endothermic)	Above 674	1245
(b) $CO + H_2O_{(v)} \rightarrow CO_2 + H_2$	+ 7.84 (exothermic)	Below 827	1520
(c) $C + 2H_2 \rightarrow CH_4$	+21.85 (exothermic)	Below 546	1014

These and other data show that above 1500°F (815°C) the net reaction tends to be (a). Below 500°F (260°C) the net reaction is a combination of (a), (b), and (c) giving:

$$C + H_2O_{(v)} \quad \frac{1}{2CH_4} + \frac{1}{2CO_2} \qquad - 1.3 \frac{Kcal}{mole} \qquad (15)$$

This reaction is almost neutral, but too slow to be practical in producing methane. Catalytic methanation is required in order to achieve reasonable production rates. The methanation reactions studied so far include:

Fixed bed (cooled either by gas recycle or heat exchange surfaces).

Fluidized bed.

Tube wall (catalyst supported on the walls of tube reactors).

The most effective catalysts found so far are nickel (especially as Raney nickel), and iron. Sulfur poisons nickel and H_2S

must, therefore, be eliminated from the gasifier product stream before introduction into the methanation reactor if nickel is used. Several processes are available for dusts, tar and H_2S removal including the Rectisol, Purisol, and Selexol systems.

Temperature for methanation should be as high as possible without carbon deposition on the catalyst, that occurs above about 850°F (454°C). Figure 6.1 is a block diagram of the SNG production process. Figure 6.2 is a schematic of the Synthane plant, one of the advanced second generation designs. It indicates one specific configuration of processes identified in Figure 6.1.

As Figure 6.1 shows, salable byproducts are generated by SNG plants in addition to pipeline quality gas. These include tar, tar oil, naphtha, crude phenol, sulfur, and ammonia. Table 6.2 provides the byproduct quantities anticipated from the 288×10^6 ft^3/day El Paso Gas design.

6.2.2 Efficiency of Operation

System efficiencies are determined by a number of variables including 1) quantity of CH_4 in the synthesis gas, 2) relative percentages of H_2 and CO in the synthesis gas, and 3) total volume of CO_2 and H_2S in the synthesis gas. Increasing the quantity of methane in the synthesis gas decreases the amount of product necessitating further treatment. As such, increasing the methane

TABLE 6.2 Anticipated Byproducts From the El Paso Natural Gas Design [7]

Product	Quantity 10^3 gal/day)
Tar	239.25
Tar oil	157.37
Naphtha	74.9
Crude phenol	32.47
Ammonia solution	332.55
Sulfur	167.0 'tons

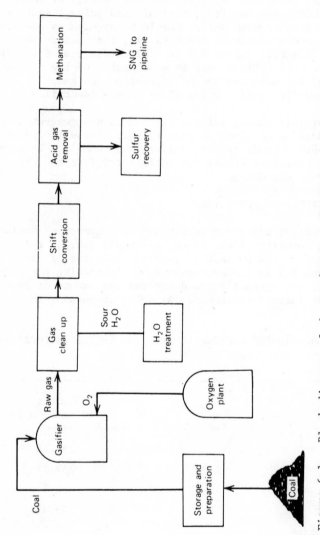

Figure 6.1. Block diagram of the production of substitute natural gas. This illustration is a generic diagram of current SNG production system. It is significant to note its similarity with Figure 4.1, the Lurgi schematic presented previously.

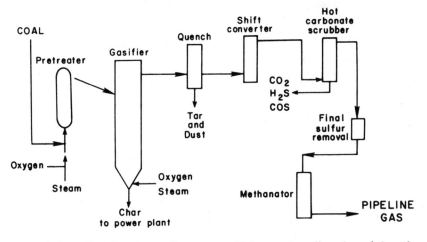

Figure 6.2. The Synthane Process. This system developed by the Pittsburgh Energy Technology Center is distinguished by its simplicity relative to other SNG systems.

content decreases water-gas shift and secondary methanation requirements and decreases the quantity of waste heat generated. The second variable, the volume ratio of H_2 and CO (ideally 3:1) again affects the water-gas shift requirement and waste heat production. It also affects process water consumption associated with the shift reaction. The CO_2 and H_2S contents affect the level of effort required in the purification steps. This variable is more influenced by coal composition than gasifier design, particularly with respect to H_2S quantities.

As Table 6.1 shows, there is wide variation among the gasifiers in approaching these variables. As a generalization, the principal distinguishing feature of first generation systems (Lurgi, Koppers-Totzek, Texaco, Winkler) is their relatively low CH_4 output. Only Lurgi produces a synthesis gas with significant amounts of methane. Second generation units consistently produce $\sim$15 percent CH_4. Comparisons between first and second generation systems on the $CO:H_2$ ratio and the CO_2 and H_2S contents are less stark. It should be noted, however, that the mean $H_2:CO$ ratio for first-generation systems is 1.2:1, and all but the Lurgi are near or below 1:1. The mean ratio for second-generation systems is 2.2:1, with no system below 1:1.

Thermal efficiencies have been calculated for several of the commercial and pilot plant systems for producing SNG. They tend to show a $\sim$58 percent efficiency for the systems operating on

methane deficient synthesis gas, ∿63 percent efficiency for the
second-generation technologies such as HyGas and Synthane [8].

In response to the idealized conditions suggested by the
efficiency criteria, Exxon Corporation has designed a highly simplified process depicted in Figure 6.3. It employs no oxygen
generating plant shift converter, or methanator. By catalytic
gasification at 1500 to 1700°F (800 to 925°C), over 20 percent
CH_4 is produced in the initial gasification. As Figure 6.3
shows, this is separated cryogenically and the $CO + H_2$ produced
are recycled to the gasifier for more methane production [2, 9].
Catalysts employed are alkali oxides. This process suggests
additional operating and thermal efficiency advantages over existing and second-generation systems.

As is obvious, then, the production tradeoff between industrial gas and SNG is one of efficiency and relative simplicity on
the one hand versus producing a storable, transportable and universal fuel on the other hand. The extent to which systems such
as Synthane or Exxon maximize methane production in the initial
gasification is the extent to which they overcome the almost
classic tradeoff that now exists.

6.3 FUEL USE CHARACTERISTICS

There is little point in dwelling on the user characteristics of SNG since, when received by the customer, it behaves as
natural gas. Table 6.3 gives some of the significant characteristics for comparison with fuels described elsewhere in this
book. Clearly this is a superior fuel carrying neither diluents
nor pollutants to the firebox.

Because SNG is a desirable product, its production from coal
has become the focal point for a considerable portion of Federal
energy research. Given the presence of over 40 million residential customers anxious for assured supplies of pipeline gas,
such a research priority is to be anticipated. Further, some 14
commercial plants capable of producing 1.2×10^{15} Btu/yr have
been proposed. The lack of appropriate Federal response to
declining natural gas reserves is an important factor that keeps
such systems from being built. The policy of the U.S. Government
has, in recent years, apparently been to buy additional natural
gas in the form of LNG (from Algeria and other countries), and
to bring in natural gas as available from Mexico, Canada, and
Alaska at prices above those allowed for domestically produced
sources (in the contiguous 48 states). Decontrol is now being
phased in ·slowly. If high Btu gas could be priced at the marginal
cost of production regardless of source (e.g., tight sandstone

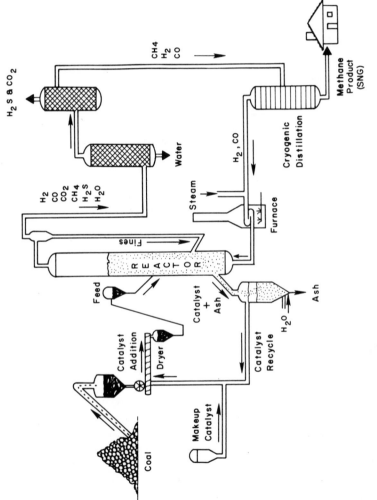

Figure 6.3. The Exxon Catalytic Gasification Process. This is another simplified system. Significant to note is the absence of an oxygen plant, a shift converter and a methanator. This is one of the most advanced concepts currently under development.

101

TABLE 6.3. Use Characteristics of Substitute Natural Gas [5]

Use Characteristic	Value
Lower heating value (Btu/ft^3)	910
Combustion air requirement (scf/10^6 Btu)	10,500
Flame temperature (°F)	3,407 (1875°C)
Volume of combustion product (scf/10^6 Btu)	11,600
Heat release/volume of combustion products (Btu/scf)	86

formations, geopressurized brines, coal gasification), then SNG from coal would have much more favorable chances for commercial deployment.

The concept of marginal pricing includes permitting utilities to obtain an attractive rate of return on SNG investments, and to use such rates as inducements for obtaining capital resources. Such rates are part of the marginal cost. Without such rates the capital markets will not support High Btu coal gasification.

REFERENCES

1 G. A. Mills, "Alternate Fuels From Coal," Chem. Tech., pp. 418-423, July 1977.

2 G. A. Mills, "Synthetic Fuels From Coal: Can Research Make Them Complete," presented before the Washington, D.C., Coal Club, March 16, 1977.

3 P. Sabatier and J. B. Senerene, C. R. Academy of Science, Vol. 134, pp. 514, 1902.

4 A. L. Hausberger, C. Bert Knight, and Kenton Atwood, "Development of Methanation Catalysts for the Synthetic Natural Gas Processes," in Methanation of Synthesis Gas, Len Seglin (Ed.), Adv. in Chem., Ser. 146, ACS, Chapter 35, 1975.

5 H. C. Hottel and J. B. Howard, New Energy Technology—Some Facts and Assessments, MIT Press, p. 103, 1971.

6 Personal communication from Bjarne R. Kristense, Rust Engineering Co., Aug. 5, 1975.

7 "Basis for Projections: Supply/Delivery Panel, Coal Conversion Group," Background paper for the Committee on Nuclear
   and Alternative Energy Systems, National Academy of Sciences,
   1977.

8 I. Howard Smith and G. J. Werner, <u>Coal Conversion Technology</u>,
   Park Ridge, N.J.: Noyes Data Corp., 1976.

CHAPTER 7

HEAVY LIQUIDS FROM COAL

7.1 INTRODUCTION

Liquefaction of coals, i.e., processing of coal where the liquid products are the main product, has been known for many years. The first work by Berguis and subsequent development of the process of coal hydrogenation have led to a variety of process conditions for producing liquids. A distinction should be made between "primary" liquefaction and other methods of producing liquid products from coal. For the purposes of this discussion "primary" liquefaction consists of one step processing where the products of the reaction between coal and other reactants is not further upgraded by secondary (or deliberately subsequent) reactions.

Several processes have been studied at laboratory and even pilot plant scale, but with liquid as a minor product. While such process may become commercial, the production of liquid is likely not to be the determining factor. One such process is the COGAS process. In this process (mentioned in Chapter 6 since SNG comprises 67 percent of the fuel product) liquids produced as fuel oil and naphtha make up only about 20 percent of the feed coal.

7.1.1 Purposes of Coal Liquefaction

The production of liquids from coal where those liquids are burned as a fuel without further processing has received increased attention in the United States as a result of the imposition of air quality standards to stationary combustion plants. The approach in many cases is to produce a liquid product low in

104

sulfur and ash, but with little other processing in order to keep the costs down. Analysis and characterization of liquid products from coal conversion by several processes reveal that the atomic H/C ratio is about the same as that for the coal feed. This indicates that hydrogen used in processing is not utilitzed in the heavy liquids to any appreciable extent. It would seem that hydrogen is not necessary at all in processing coal to heavy liquids; however, in a practical sense this has not proved to be the case. Hydrogen is used to react with sulfur and oxygen especially in terminal structures and bridges in the coal. Sulfur in inorganic forms is virtually completely removed by reaction with hydroben, producing H_2S. Reaction of organic sulfur in ring structures, such as thiophene, is not accomplished while organic sulfides, disulfides, and some other organic sulfur-containing compounds are desulfurized. The result is that sulfur elimination from the heavy liquid product depends on the relative amounts of inorganic and organic sulfur present in the coal. In most cases where inorganic (pyritic) sulfur constitutes about one half of the total sulfur in the coal 70 to 80 percent can be removed by primary liquefaction.

Liquefaction, in this country, permits the production of a storable, transportable substitute for petroleum. Within the confines of coal conversion it offers the only relatively ubiquitous alternative to SNG. Outside of the United States different reasons for coal liquefaction may be present. In the United Kingdom, for instance, the interests in coal liquefaction have been given as production of high grade products (i.e., carbon for electrodes and carbon fibers) and production of feedstocks for the manufacture of synthetic hydrocarbons [4]). In this chapter consideration is for heavy liquids, that is, the liquid product from solvent treatment or other liquefaction where sulfur removal and separation from inorganic constitutents of the coal are the main objectives. Refining or upgrading are not considered since light liquids and chemicals from coal are treated in a subsequent chapter.

7.1.2 Conditions Necessary for Liquefaction

The conditions necessary for the liquefaction of coal include temperatures above the softening temperature (325 to 375°C for bituminous coals). Under such conditions the following processes may be used to produce liquids.

1. Reaction with hydrogen under pressure.
2. Pyrolysis (carbonization).
3. Hydrocarbonization (combination of 1 and 2).
4. Solvent extraction (dissolution or solvent refining).

The conditions required for liquefaction of coals also produce gases and solids. In some cases the process may be designed to produce liquids which are higher in their atomic H/C ratio than the original coal. These are not considered as "primary" liquefaction processes since the product has been deliberately upgraded by secondary reactions to produce a distillate.

The production of liquids from coal may have several purposes. Even the term liquid has to be defined since some products of coal liquefaction may be solids at ambient temperatures. For the purposes of this chapter heavy liquids will include all material soluble in some organic solvent used to measure the conversion of the process. Many solvents have been used to measure conversion. For a long time benzene solubility in a soxhlet apparatus was used to define conversion [1]. Solvents used to define conversion included pyridine [2] quinoline [3] as well as other compounds and mixtures.

Traditional characterization of coal-derived liquids by the German workers resulted in definitions of coal liquefaction products that served their purposes, but not those of today when more sophisticated analytical methods are available. Table 7.1 lists terms and their definitions from both past and current work. The overlapping of definitions makes comparisons difficult especially when different coals have been used and the methods of reporting conversion are not the same.

7.2   PROCESSES FOR COAL LIQUEFACTION

Utilizing the methods listed in the previous section many processes have been used for liquefaction of coal. Not many of these have been carried out at full scale production levels. The exceptions are Germany during World War II and South Africa presently. For the Germans economics had to take a secondary position relative to national objectives. A similar situation, although not a world war, led to synthetic liquids production in South Africa. The small scale of operation used in Germany (by today's standards) and high costs predicted for such processing prevented liquefaction of coal in most of the world. Impetus for coal liquefaction has been enhanced by the dramatic increase in crude oil prices of 1973 and considerations of national security. South Africa and the United States both have adequate reserves of coal, but could be blackmailed by an embargo of crude oil. European countries are in a somewhat similar situation but their coal prices make processing to petroleum substitutes less attractive than for such countries as the United States.

Because of the increased interest in coal liquefaction at the beginning of the 1960s, there has been considerable evaluation

TABLE 7.1 Definition of Terms Used in Coal-Derived Liquid Characterization

Term	Other Parallel Terms	Definition[a]
Oil		Material soluble in pentane (hexane, heptane, and cyclohexane have also been used).
Asphaltenes		Matter soluble in benzene, but insoluble in pentane.
Asphaltols	Preasphaltenes	Matter soluble in pyridine, but insoluble in benzene.
Solvent Refined Coal (SRC)		Organic material in coal that has undergone dissolution or suspension in a solvent (usually at T>375°C). During the processing most inorganic material in the coal has been removed; also the solvent has been at least partially removed. SRC is a solid at room temperature the softening point (T) depending on the degree of solvent removal.
Residue	Char, coke, unreacted coal, prompt residue	Material insoluble in benzene or pyridine.
Pitch	Coal tar pitch	The residue left after coal tar is redistilled (usually to a final temperature over 400°C, the exact endpoint depending on the pitch desired).

[a]Solubility is usually measured at or near the atmospheric boiling point of the solvent involved. The usual procedure is by use of a soxhlet extraction apparatus with extraction carried out until clean solvent is recycled.

of the prospects of coal liquefaction by pyrolysis, direct hydrogenation, solvent extraction (and refining), and by gasification
to synthesis gas with subsequent liquid production. Research
work as well as bench scale and pilot plant operations were conducted by the Bureau of Mines since before World War II. In 1962
the Office of Coal Research began funding research and development work on conversion of coal up to and including pilot plant
operation. These programs, subsequently absorbed into ERDA and
the U.S. Department of Energy, have also participated with industry in process development and demonstration. Table 7.2 lists
pilot plants built for such purposes and the method of conversion
employed in each.

7.2.1 Conventional Conversions and Products

1. Low Temperature Tar (LTC). Outside of Germany and England
little commercial production of low temperature (less than 930°F
or 500°C) tar has taken place, although significant effort has
been made to develop processes. The price advantages of petroleum has prevented large scale development. Low temperature tar
(LTC) has importance both historically and theoretically as far
as the coal liquefaction and chemical composition are concerned.
The character of the tar produced by low temperature carbonization depends on the nature of the coal processed. The most uniform characteristics of LTC are the low concentrations of any one
chemical compound (usually less than 2 percent) and the relation
between carbonization temperature and the quantity of unalkylated
aromatic material.

LTC's contain low boiling components such as phenols, alkyl
benzenes, n-paraffins, naphthenes isoparaffins, and alkylthio-
phenes [5] along with many other aromatic hydrocarbons [6]. Com-
pilations of compounds that have been identified along with their
amounts for both distillates and heavy ends have been made by
Karr [7] and the reader is referred to these for more specific
chemical information.

From earlier studies of the composition of LTC's and the conditions used for their production it has been noted that lower
carbonization temperatures and shorter residence times produce
tars with significantly low concentrations of aromatics. The
major components of low temperature processing (T<840°F) were
alkylated naphthalenes [8] and aliphatics [9] with aromatics
becoming more evident as temperatures were increased above 840°F
(450°C).

The major utilization of LTC has been as liquid fuels essentially as a synthetic crude petroleum. Thus fuel oils, diesel
oils and even gasoline have been produced. Although the primary
use has not been chemicals considerable quantities of phenols and

TABLE 7.2 Pilot Plants for Coal Liquefaction

Plants and Location	Sponsor(s)	Capacity (tons/day)	Conversion Methods	Date
Project-Gasoline Cresap, WV	Office of Coal Research/ Consolidation Coal Company	20	extraction + catalytic hydrogenation	1967–1970
COED, Princeton, NJ	Office of Coal Research/ FMC	36	staged pyrolysis	1970–1975
H-Coal, Trenton, NJ	HRI, ARCO, Ashland Oil, Esso, Amoco	3	hydrogenation in ebullated catalyst bed	1968–
Solvent Refining (PAMCO), Ft. Lewis, WA	Gulf (PAMCO)/ERDA	50	dissolution + filtration	1974–
Solvent Refining Wilsonville, Ala.	Southern Services	6-12	dissolution + filtration	1974–
Exxon Donor Solvent	Exxon	1	extraction-hydrogenation	1975–

cresylic acids have been produced and used for the manufacture of resins and other products. LTC has also been used for road tar, mastic for flooring, fungicides and wood preservatives, flotation agents, and dyes.

Because no commercialization of low temperature carbonization exists in the United States, at present the economics of LTC production must be considered as unfavorable, although the increasing price of petroleum may change this situation. The commercial production of LTC would appear to be a definite probability in the future, the exact time for the beginning of such production to be determined by crude oil and coal prices. In comparison to the production of other liquids from coal LTC would seem to be one of the most economically feasible due to the relatively mild conditions of temperature and the fact that no pressurized equipment, hydrogen source, or solvent handling is involved.

2. High Temperature Tar. High temperature tar (HTC) is produced from coal by destructive distillation of coal where final temperatures exceed 1470°F (800°C). Although many different designs for retorts are and have been used for such distillation (carbonization) the major portion today are by-product metallurgical coke ovens, especially in the United States.

In the carbonization process at high temperatures HTC is produced as well as "light oil" and gases. The high temperature carbonization of coal in coke ovens is optimized for the production of suitable coke properties and yield since this is the primary product. During the carbonization (pyrolysis) reactions many cracking and elimination reactions occur that break bonds in the structures comprising coal and primary products described in the previous section on LTC. Such reactions result in the formation of many light distillable products such as BTX (benzene, toluene, and xylenes), other light aromatics and saturated hydrocarbons (very low in concentration) and derivatives, water, H_2S, CO_2, H_2, CH_4, and other hydrocarbon gases and ammonia.

The loss of hydrogen, aliphatic, and hydroaromatic materials results in a high degree of aromatization in the liquids and solid coke not volatilized by such reactions. The resulting HTC, not distillable, contains condensed aromatic ring structures having more than five rings. In the limit, if high temperature carbonization is continued, all hydrocarbons would be converted to hydrogen and graphite; the HTC from coke ovens or other high temperature carbonization are some intermediate of this ultimate process with the nature of the HTC depending on the carbonization time and the final temperature reached during the process.

Compositions of HTC and coal tar pitch produced from high temperature carbonization have been reviewed by Weiler [10] with

information on the influence of the coals used, the final temperatures and the particular process equipment employed [11-13].   In
view of this and other data available only general information is
given on HTC characteristics and uses.

Whereas LTC composition may depend considerably on the coal
used for its formation, HTC composition depends mostly on the
final temperature achieved during carbonization.   HTC's also contain greater percentages of aromatic compounds and a specific
compound may be present to the extent of several percent (naphthalene, for example, has been found to constitute 10 percent of
some coke oven tars [7]).   As indicated earlier the composition
of HTC is variable and depends on the conditions of carbonization
and to a lesser extent on the properties of the feed coal.   The
following is a representative description of HTC produced with a
final carbonization temperature of 2000°F (1100°C)

	Volume Percent
Neutral oils	18.0
Tar acids	2.5
Tar bases	1.0
Free carbon	13.0
Specific gravity, (g/ml)	1.20

Significant quantities of coal tar pitch and water are also in
the HTC as produced.   Water is particularly troublesome since it
causes foaming in distilling operations, detracts from the heating value of the tar, and can promote corrosion of metals con-
tacted by wet tars.   Moisture is usually removed by dehydration
with heating of the tar and steam together although centrifugation has also been used.   Tars sold commercially usually must
contain less than 2 percent water.   HTC is usually separated and
refined to produce both pure (or relatively pure) products such
as BTX and naphthalene, and crude fractions such as creosote oil,
pitches, tar acids, anthracene oil, and pyridine bases.   The
crude fractions are not of definite composition and combinations
and subfractions are sometimes prepared.   When these materials
are used as fuels instead of chemicals, physical properties such
as heat of vaporization, thermal conductivity, viscosity, and
specific gravity are more important than chemical types present.
In the case of the heavy tar and pitch fractions the common
method of characterization is by viscosity and viscosity-
temperature behavior.   These materials are typically non-
Newtonian and emperical equations are used to describe or predict
viscosity temperature relationships.

HTC processing has been shown to give best returns by separating tar acids and tar bases as well as recovering naphthalene, creosote oil and in some cases other chemicals before ending up with heavy tar and pitch. Refined tars are also extensively used as roofing material, metal pipe coverings for the prevention of underground corrosion, preservatives, and road tar. In these and other uses coal tar pitch is superior to other available materials due to its high resistance to water and other aqueous solutions. Coal tar pitch has many other uses including carbon electrode and silicon carbon manufacture and as a fuel. Since HTC and its fractions are low in ash and sulfur, these can be desirable fuels especially in the steel industry. The heating value of these heavy fractions is usually at least 16,000 Btu/lb.

7.2.2 Recently Developed Processes for Heavy Liquid Fuels From Coal

1. Solvent Refined Coal (SRC). Many early experiments were conducted to determine the solubility of coal in various solvents not only to study coal and its reactions, but also as a possible method of separating the valuable fuel portion of coal from its inorganic impurities. The Pott-Broche process was successful in dissolving about 75 percent of the German bituminous coals processed and producing an extract with only 0.5 percent ash [14]. Plants were operated in Germany in World War II until war damage prevented further processing. Japan also operated such a plant in World War II [15]. The main use of the German process was to produce a low ash coke for aluminum manufacture. A similar type of product is the objective of coal liquefaction in England where graphite electrodes are produced.

The restriction of sulfur oxide emissions from stationary combustion sources in the United States has been largely responsible for development of a process for production of a low sulfur, low ash coal. This solvent refined coal (SRC) is obtained by dissolving, suspending or liquefying the coal in a high boiling (550 to 800°F, 300 to 425°C) coal derived solvent in a manner similar to the Pott-Broche process. Many companies, electric power groups, and the government have been involved in the development and evaluation of SRC production.

a. Production Processes. Research in the 1950s was performed by Spencer Chemical Company on a variation of the Pott-Broche process to produce a clean heavy liquid or solid from coal. Under sponsorship of the Office of Coal Research (U.S. Department of Interior) a 50 lb (23 kg)/hr pilot plant was built and operated by Spencer Chemical Company in Kansas City to evaluate the process [16, 17]. Spencer Chemical Company workers

now part of Gulf Research and Development's subsidiary Pittsburgh
and Midway Coal Mining Company (PAMCO) and others have done considerable testing and development of the process and of SRC properties. As indicated in Table 7.2 two pilot plants were built to
obtain engineering data and to produce sufficient quantities of
SRC for product evaluation. The essential features of SRC production are shown in Figure 7.1.

In the SRC process coal is ground 0-18", dried to <3 percent
moisture and slurried with solvent and hydrogen before introduction into the preheater. The mixture is reacted in the dissolver
at up to 800°F (425°C) and 2000 psi where over 90 percent of the
organic portion of the coal is dissolved or suspended. Some
chemical reactions are involved under such conditions. These may
include 1) thermal and catalytic (utilizing minerals in the coal)
bond rupture, 2) hydrogenation of reactive fragments from 1),
3) solvent dissociation, dehydrogenation and isomerization, and
4) removal of oxygen and sulfur from the coal and inorganic materials associated with the coal. This latter step usually
reduces the concentration of sulfur to 10 to 15 percent of its
original concentration in the coal. Since minerals are not soluble in the organic solvent they may be removed by filtration,
centrifugation or other methods. The extract is flash distilled
to separate solvent.

The SRC product may be recovered hot as a pumpable liquid or
cooled to a low sulfur, low ash solid. The softening point (temperature) of the product SRC depends partially on the degree of
solvent removal. SRC produced from a variety of coal feeds has
essentially the same physical gross properties. The ash content
is low (0.1 to 0.5 percent) and the heating value is about 16,000
But/lb (8900 cal/gm). H_2S, CO_2, and some fuel gases are produced
in the process.

Since the work done by Spencer Chemical, now PAMCO of Gulf,
others have also worked on development of the processing of coal
to SRC. These include Old Ben Coal Company, Mitsui and Company,
Combustion Engineering and Catalytic, Inc. [15]. In addition,
many SRC-related studies have been and are being done by various
universities, engineering companies, and research institutes
[15, 18-20]. The primary purpose for production of SRC has been
to make available a low pollution material suitable for combustion in electric power plants. Other uses may be found for SRC
as larger quantities become available from commercial production.

The most difficult problems associated with SRC production
concern the liquid-solids separation step. Although pressure
filtration, centrifugation, and some other methods have been suggested and studied commercial experience determines the best separation method.

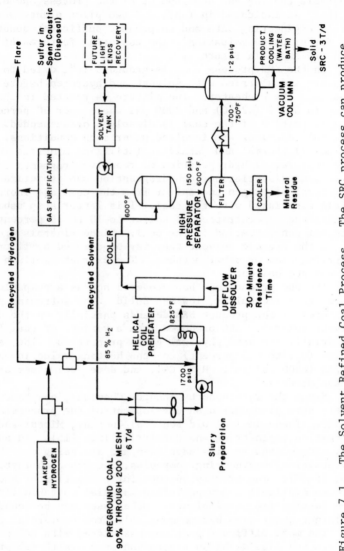

Figure 7.1. The Solvent Refined Coal Process. The SRC process can produce both clean solid coal fuels and heavy liquid boiler fuels.

Because of the difficulties mentioned in the solid/liquid separation, development of SRC-2 was pursued by Gulf. The basic idea behind this development was to produce a distillable liquid product. This would eliminate the need for filters, centrifuges, and other physical separation methods, that Gulf spent consider- able time and money studying and testing. The penalty for pro- ducing a distillate product is approximately a doubling of the hydrogen required in the process. While SRC production requires about 2 percent hydrogen (based on the feed coal), the SRC-2 requirement is 3.5 to 4.0 percent hydrogen. However, the SRC-2 product can be separated from residue and mineral matter by flash vaporization or distillation.

b. Fuel Characteristics. Since pilot plant experience has been obtained for SRC production, considerable quantities have been tested for combustion behavior. Although insufficient data is available on the specific properties affected by different coal feeds, it appears that solvent refining produces a somewhat uniform product regardless of the feed coal, at least as far as the gross properties are concerned. Reports on combustion char- acteristics report the following as representative of SRC [21].

Ash	$\sim$0.1 percent
Sulfur	0.5 to 0.9 percent
Volatile matter	$\sim$60 percent
Nitrogen	$\sim$2.0 percent
Heating value	18,800 to 16,000 Btu/lb (8800 to 8900 cal/gm)

Other important combustion properties of SRC include igni- tion data, carbon residue, grindability index, and trace miner- als. Some of these depend on the nature of the feed coal. For a Kentucky high volatile bituminous coal from the Wilsonville, Alabama pilot plant the following were found:

i. SRC had a higher pour point, Conradson carbon residue and flash and fire points than a No. 6 fuel oil.
ii. Ignition temperature was found to be similar to a high volatile bituminous coal.
iii. Burnout temperature was found to be similar to a higher rank coal such as a low volatile bituminous or anthracite coal.
iv. Hardgrove Grindability Index was relatively high ($\sim$180) and was constant below about 150°F (66°C). Above this

temperature grinding of SRC was more difficult and as
the SRC became tacky near 280°F (140°C), grinding was
not possible.

v.   SRC was found to be "considerably more viscous" than
No. 6 fuel oil. SRC also undergoes the change in
viscosity characteristic of coal-derived liquids when
exposed to air [22, 23], a characteristic caused by
oxidation.

vi.   SRC was found to agglomerate easily, which resulted in
serious feeding problems if the SRC was stored before
feeding.

In general the prospects for combustion of SRC are excellent.
The low ash and lower sulfur content give SRC low pollution potential. No visible plumes were observed from SRC firing except
when inadequate excess air was used. With SRC of 0.9 percent
sulfur SO_2 emissions were relatively low (500 ppm) compared to
2200 ppm for combustion of the parent coal. The most serious
problems were feeding problems (that could be solved by in-line
pulverizers, to avoid agglomeration) and NO emissions. Since the
nitrogen content of the SRC was relatively high the NO emissions
were about 200 ppm higher for combustion of the SRC than for the
parent coal. SRC can be used as a liquid fuel, but requires
heating to 410°F (210°C) in order for it to be pumped as a liquid;
additionally, atomization requires temperatures in excess of 540°F
(280°C). Because of the oxidation reactions of liquid SRC with
air, storage as a liquid would require inert gas blanketing.

2. Other Coal-Derived Liquid Fuels. Many bench scale, process
development, and pilot projects for liquefaction of coal to produce clean liquid fuels have been undertaken in recent years.
Although these are not discussed in detail some points should be
made regarding them as pertaining to heavy liquids from coal.
The properties of coal derived liquids with the same hydrogen and
oxygen content and H/C ratio have similar characteristics as far
as combustion behavior and gross physical properties. Programs
to develop processes for conversion of coal to clean liquids
utilize one of the methods listed at the beginning of this chapter namely; hydrogenation at high pressure, pyrolysis, hydrocarbonization, or solvent extraction. Additionally gasification of
coal may be performed with subsequent synthesis of the H_2 and CO
to produce hydrocarbon liquids or other synthesis products
(Fischer-Tropsch synthesis).

In addition to the pilot plants listed earlier in this chapter, smaller scale units (but larger than bench scale) are, or
have been, operated by Universal Oil Products, Continental Oil Co.,

Hydrocarbon Research, Conoco Coal Development Co., Lummus, University of Utah, North Dakota University, and ERDA (Pittsburgh
Energy Research Center, now U.S. Department of Energy).   Two of
these are especially interesting since they do not employ the
usual methods of processing coal to produce liquids; the Department of Energy work (Synthoil) and the University of Utah
process [24].

In most liquefaction processing the coal is either suspended
in a vehicle oil to make a slurry or dissolved in a solvent.   The
duration of time that this liquid or slurry is at the process
conditions has varied from a few minutes to an hour or longer. In
the Synthoil process shown in Figure 7.2, a coal/oil slurry is
fed to a reactor packed with a fixed bed catalyst.   The slurry
along with hydrogen travels through the reactor under turbulent
conditions where contact between the catalyst, coal and carrier
(coal derived liquid), and hydrogen produce hydrogenation and
desulfurization reactions.   The shorter residence times (a few
minutes or less) and turbulent contact with the catalyst (cobalt
molybdate) bed make this process somewhat unique.   The Synthoil
product although a liquid at room temperature is a premium fuel,
with many similarities to SRC.   Some extensive analyses have been
made on Synthoil products including the nature of the sulfur
compounds [27].

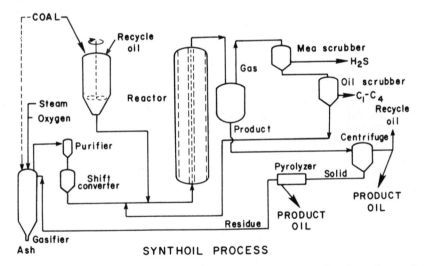

SYNTHOIL PROCESS

Figure 7.2. The Synthoil Process. This system, developed at the
Pittsburgh Energy Technology Center (DOE), is one of the leading
liquefaction systems under development. It is distinguished by
short residence times in the reactor and the use of a fixed bed
catalyst.

The University of Utah process shown in Fig. 7.3 is unique in that dry powdered coal is fed, along with hydrogen, to a coiled tube reactor. The residence time in the reactor is even less than with Synthoil (5 to 240 sec), with direct measurement of the residence time accomplished [25]. A volatile catalyst ($ZnCl_2$) is mixed or impregnated into the feed coal before its introduction to the reaction system. Turbulent flow is essential for suspension of the solid coal and catalyst as well as to maintain continuous operation of the system [26]. The ratio of liquids to gases in the products can be quite high (6 to 7:1).

7.3 PROSPECTS FOR COMMERCIALIZATION

Commercialization of liquefaction processes, although not now a reality, is a virtual certainty as petroleum supplies diminish while environmental standards become increasingly stringent. Such commercialization is inevitable because it meets the energy needs of the following users:

1. Small and medium sized manufacturers located away from major deposits of coal and away from natural gas (with SNG) pipelines.
2. Manufacturers that do not have sufficient real estate available to support coal storage, combustion or conversion facilities.
3. Utilities that, for one reason or another, cannot use coal directly and cannot justify coal conversion as part of their activities.

Already construction has begun on a large (600 TPD) pilot plant in Catlettsburg, Kentucky, a plant scheduled to be producing 1800 bbl/day by mid 1983.[*] Wheelabrator Frye is projecting construction of a 20,000 TPD SRC plant on a similar timetable.[†] Various studies, including that of the Committee on Nuclear and Alternative Energy Systems (CONAES) of the National Academy of Sciences, have forecast at least one commercial (50,000 bbl/day) refinery by 1985 and rapid growth thereafter.

While the economics of heavy liquid production are discussed in Chapter 9, it is clear that first, second, and third generation plants will be evolving to serve the industrial markets.

---

[*]"Largest Pilot Plant Started for Coal-Liquefaction Test," Energy User News, Dec. 27, 1976, p. 10.

[†]Eugene Ward, "Non-Polluting Coal is Wheelabrator Aim," Energy User News, Aug. 30, 1976, p. 17.

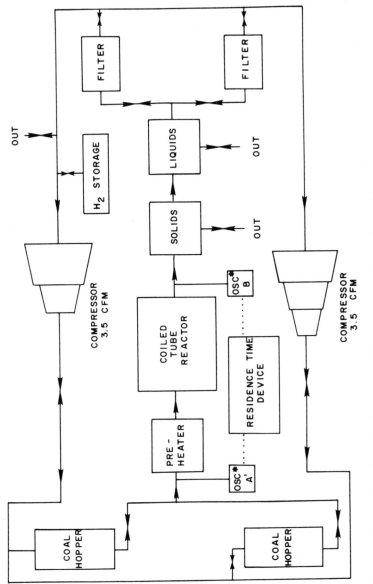

Figure 7.3. The University of Utah Process. This liquefaction system is distinguished by its feeding of dry coal into the reactor. Other systems slurry the coal in a liquid medium for feeding purposes.

Many process development projects and evaluations have been made to determine the technical and economic feasibility of coal liquefaction to produce clean liquid boiler fuels without the expense of producing distillate fuels or chemicals. The production of the latter light liquid fuels and chemical feedstocks, as discussed in the following chapter, requires more extensive hydrogenation and often two stage reaction systems.

REFERENCES

1 W. R. K. Wu and H. H. Storch, U.S. Bureau of Mines, Bull. 633, 1968.

2 D. D. Whitehurst and T. O. Mitchell, Amer. Chem. Soc.-Fuel Chem. Div., Preprints, Vol. 21, No. 5, p. 127, 1976.

3 D. L. Kloepper, T. F. Rogers, C. M. Wright, and W. C. Bull, U.S. Dept. of Interior, Office of Coal Res. R & D Report No. 9.

4 J. S. Harrison, Coal Processing Technology, Vol. 2, Chemical Eng. Progress Technical Manual, A. I. Chem. Engrs., pp. 20-24, 1975.

5 A. Jager and G. Kattwinkel, Brennstoff-Chemie, Vol. 35, pp. 353-362, 1954.

6 T. Chang, C. L. Chang, and C. Karr, Jr., Anal. Chim, Acta, Vol. 21, pp. 474-490, 1959.

7 C. Karr, Jr., Chemistry of Coal Utilization, Suppl. Volume, H. H. Lowry (Ed.), New York: J. Wiley and Sons, Chapter 13, pp. 544-555, 1963.

8 M. J. Vahrman, Appl. Chem. (London), Vol. 8, pp. 485-492, 1958.

9 F. Trefry, Tech. Mitt, Vol. 48, pp. 223-233, 1955.

10 J. F. Weiler, Chemistry of Coal Utilization, Suppl. Vol., H. H. Lowry (Ed.), New York: J. Wiley and Sons, Chapter 14, pp. 587-600.

11 E. O. Rhodes, Chemistry of Coal Utilization, Suppl. Vol., H. H. Lowry (Ed.), New York: J. Wiley and Sons, Chapter 31, pp. 1287-1370, 1945.

12 Coal Tar Research Association, The Coal Tar Data Book, Section B1, 1-3, Gomersal, England, 1957.

13 P. J. Wilson, Jr., and J. H. Wells, Coal, Coke and Coal Chemicals, New York: McGraw Hill Book Co., Inc., Chapter 12, pp. 372-399, 1950.

14 W. M. Sternberg (Translator), "High Pressure Hydrogenation at Ludwigshafen-Heidelberg," Vol. IA, General Section, FIAT Final Report 1317, ATI No. 92, 762, for Central Air Documents Office, Dayton, OH, 1951.

15 Battelle, Columbus Laboratories, "Liquefaction and Chemical Refining of Coal," A Battelle Energy Program Report, p. 105, 1974.

16 D. L. Kloepper, T. F. Rogers, C. H. Wright, and W. C. Bull, "Solvent Processing of Coal to Produce a De-Ashed Product," R & D Report No. 9, U. S. Dept. of Interior, Office of Coal Research, Washington, D.C., 1965.

17 Anon., "Economics of a Process to Produce Ashless, Low-Sulfur Fuel From Coal," R & D Report No. 1, U.S. Dept. of Interior, Office of Coal Research, Washington, DC, 1970.

18 R. M. Baldwin, J. O. Golden, J. H. Gary, R. L. Bain, and R. J. Long, Chem. Eng. Progr. Vol. 71, No. 4, pp. 128-129, 1975.

19 R. G. Shaver, "A Solvent Refined Coal Process for Clean Utility Fuel," Pollution-Control and Energy Needs, Adv. in Chem. Series No. 127, American Chem. Soc., Washington, DC, pp. 80-90, 1973.

20 I. Howard-Smith and G. J. Warner, "Coal Conversion Technology," Chem. Tech. Rev., No. 66, Noyes Data Corp., Park Ridge, NJ., pp. 33-34, 1976.

21 Anon., "Investigating the Storage, Handling and Combustion Characteristics of Solvent Refined Coal," EPRI 1235-1, Final Report (prepared by Babcock and Wilcox Company), 1975.

22 Y. Y. Lin, L. L. Anderson, and W. H. Wiser, Amer. Chem. Soc. Fuel Chem. Div. Preprints, Vol. 19, No. 5, pp. 2-32, 1974.

23 F. S. Karn, F. R. Brown, and A. G. Sharkey, Jr., Amer. Chem. Soc., Fuel Chem. Div. Preprints, Vol. 22, No. 2, pp. 227-232, 1977.

24 S. B. Alpert and R. M. Lundberg, in Annual Review of Energy, J. M. Hollander (Ed.), Annual Reviews, Inc., Palo Alto, CA, Vol. 1, pp. 87-99, 1976.

25 J. M. Lytle, R. E. Wood, and M. G. Mladejovsky, Fuel Proc. Tech., Vol. I, No. 2, pp. 95-102, 1978.

26 L. L. Anderson, R. E. Wood, and W. H. Wiser, Trans. AIME Soc. Min. Eng., Vol. 260, No. 4, pp. 318-321, 1976.

27 S. Akhtar, A. G. Sharkey, Jr., J. L. Schultz, and P. M. Yavorsky, Amer. Chem. Soc.-Fuel Chem. Div. Preprints, Vol. 19, No. 1, pp. 207-214, 1974.

CHAPTER 8

LIGHT LIQUIDS
AND CHEMICALS FROM COAL

8.1 INTRODUCTION

Due to the relatively simple transformation and the minimal hydrogen requirement for solvent refining of coal or conversion to heavy liquids, the production of light products is considered by some to be less efficient processing for coal. However, light liquids that can be used as motor fuels or processed to valuable chemicals are increasing in use and demand. Recent studies also show that increased quantities of petroleum and natural gas will be required to satisfy future demands for petrochemical feedstocks [1]. In fact there is considerable evidence that it may become necessary to obtain such feedstocks from nonpetroleum sources such as coal and shale oil. In the past coal has been a significant source of some chemicals, primarily as by-products from carbonization in by-product coke ovens. Other examples of processes for the production of distillable liquids from coal include:

Gasification to synthesis gas (H_2 + CO) followed by Fischer-Tropsch synthesis.

Plasma pyrolysis of coal at high temperatures.

Fractionation of Solvent Refined Coal (SRC) products or other "primary" liquefaction processes, including solvent extraction, hydrogenation, etc.

Hydrocarbonization (vis. "Coalcon" process) [2].

Since light oils (distillation range 60 to 150°C) are more amenable to detailed analysis than heavier liquids, considerable

data are available on the composition of such liquids. Light oil
has been the chief source of some chemicals in the past and will
continue to become more important in this respect. Chemicals
that have been produced from coal as a significant source have
included: ammonia, benzene-toluene-and xylene (BTX), naphthalene, phenanthrene, phenol, ammonium sulfate, sulfur, carbon dioxide (dry ice and liquid), anthracene, and tar bases.

    In many cases the position of coal in supplying chemical
materials has been eroded due to the lack of large scale processing of coal. Two examples are ammonia and benzene. Ammonium
sulfate is produced as a by-product of coke production as is
benzene. Synthetic ammonia has far exceeded the production from
coal because only one-fourth of the solid ammonium sulfate is
ammonia and the production is limited by the production of coke.
In the case of benzene, coal was the major source and continues
to supply significant quantities. However, coke-oven benzene
production was matched by petroleum sources by 1958 and has continued to increase [3]. By 1977 400,000 tons/yr of benzene were
being produced in the United States with over 80 percent coming
from petroleum.

    The discussion of light products from coal is divided into
two categories; those produced from light oil and those obtained
by other means. In the past most light oils have been produced
by carbonization in by-product coke ovens. After reviewing light
oils and the chemicals contained therein production of other
chemicals such as methanol, ammonia, and acetylene are considered.

8.2 LIGHT OIL

    Coal carbonization produces a full range of liquid and gas-
eous products as indicated in the previous chapter. Light oil is
the fraction boiling between 60° and 150°C, although some products may be termed "light oil" with slightly different distillation ranges. Crude light oil from coal carbonization is composed
of mainly (over 80 percent) one ring aromatic hydrocarbons and
their derivatives. In most cases benzene, toluene and zylenes
(BTX) comprise more than 80 percent of the total weight. Table
8.1 gives some typical BTX contents of light oils from various
sources. Many other analytical results have recently become
available for light oils that have been produced and marketed for
years without the composition having been known. The development
of gas chromatography and mass spectroscopy have made possible
detailed analysis of essentially all compounds in the light oil
range, including hydrocarbons and oxygen- and sulfur containing
compounds [5, 7]. Analyses have also been given for these light
oils by compound type [8]. The BTX compounds comprise most of

TABLE 8.1 Hydrocarbons in Crude Light Oils from Durham Coal[a]

Compound	Source of Product (% by weight)		
	Coke Oven	Horizontal Retort	Intermittent Vertical Chamber
Unsaturated Compounds			
Styrene	1.11	0.61	1.23
Indene	1.57	0.69	1.62
Cyclopentadiene and dicyclopentadine	0.14	1.79	0.49
Paraffinic	0.42	1.09	3.27
Naphthenic	0.23	0.98	2.74
Aromatic			
Benzene	72.4	68.4	45.5
Toluene	13.5	11.4	18.9
Xylenes	2.92	2.53	7.71
Total BTX	88.82	82.53	72.11
Other Aromatics	1.41	0.90	2.93
Total Product Determined as Hydrocarbons	93.70	88.59	84.39

[a]Sources: Benzole Producers Ltd., Half-Yearly Report-Technical Dept. 14, Sept. 1985. G. Claxton, Benzoles: Production and Uses, National Benzole and Allied Products Assoc., London, 979, 1961.

the light oil with the minor constitutents being olefins (9 to 10 percent), paraffins (<1 percent), sulfur compounds ($\sim$1 percent) and other compounds (3 to 4 percent).

The exact mechanism of the reactions for the high temperature carbonization of coal to light oil is not known. However, some have been proposed. We can at least improve our understanding of the carbonization process by careful observance of the products and reactants and how they are affected by coking conditions. The composition of the light oil from by-product coking is very

little affected by the coal since the final coking temperature
is above 900°C. The amount of light oil obtained per unit weight
of coal is affected by the coal, being notably higher for certain
high volatile bituminous coals. The final temperature, to which
the carbonization products are heated, seems to influence the
light oil composition more than any other variable. The most not-
able effect is that light oils from high temperature ovens is
mostly aromatic (as shown in Table 8.1) whereas products in the
low boiling range from low temperature carbonization often contain over 60 percent saturated hydrocarbons [9]. The maximum
yield of aromatics occurs below 900°C, but slightly above the
temperature range where maximum tar yield is found (500 to 700°C).

     Although explanations have been given to explain the types of
reactions leading to light oil formation most of these do not
take into account the influence of time. Recent experiments have
suggested that initial reactions of coal regardless of the temperature employed result in larger molecules than those in light
oil. This would mean that light oil is composed of secondary
products produced by cracking, hydrogenation, and splitting reactions. As reaction time is longer, more cracking and condensa-
tion reactions occur leading to condensed structures (coke) and
light oil and gas at the expense of heavy tar products. Early
explanations were based mostly on qualitative product analyses
and thermodynamic and experimental studies without the benefit of
detailed analyses of products, especially as a function of reaction time [10-12].

8.3 LIGHT LIQUIDS BY HYDROGENATION METHODS

     Light liquids products may be produced from coals by several
methods as listed at the beginning of this chapter, even those
specifically designed to only remove mineral matter and reduce
sulfur such as solvent refining. As explained above when reaction times are long (more than a few minutes) more light oils and
gases are produced at the expense of heavier liquids. This holds
for the processes referred to except for liquids produced by a
two-step process such as gasification to synthesis gas (H_2 and CO)
followed by catalytic synthesis to final products. The maximum
temperatures to which oil to tar vapor can be subjected without
cracking to gases has been given as 480 to 520°C [13]. The reactions of tar vapors above this temperature are increased as the
temperature increases and is enhanced by the presence of metal
surfaces and inorganic minerals in the coal that can serve as
catalysts.

     Attempts to produce light liquids directly from coal were
considered to be impractical even by Friedrich Bergius, the

inventor of coal hydrogenation. This was due to the evolution of
H_2S during hydrogenation reactions and the known poisoning effects
of sulfur compounds on the hydrogenation catalysts used at the
time of the discovery of coal hydrogenation (1911), namely,
platinum, palladium, and nickel.
    Commercial operation in Germany during World War II of the
Bergius process was carried out as a two-stage operation.

1. Reaction of coal in a vehicle oil accomplished by a
    catalyst. The product of this stage is a middle oil
    (distillable) that is used as the feed to the second
    stage.
2. Reaction of middle oil over a fixed-bed catalyst to
    produce gasoline-range products.

In the United States several attempts have been made to produce gasoline or naphtha materials directly from coal without
success. Recent work has been directed to the production of
light liquids by the following methods.

1. Gasification of coal to synthesis gas followed by cata-
    lytic synthesis of the CO and H_2 to methanol (CH_3OH).
    Mobile researchers have reacted methanol over a catalyst
    to produce gasoline compounds and water [14]. The Mobil
    process employs a selective catalyst with controlled
    pore sizes to produce the desired products.
2. Gasification to synthesis gas followed by catalytic syn-
    thesis of hydrocarbons (Fischer-Tropsch synthesis). By
    proper selection of catalysts and reaction conditions
    products ranging from methane to high molecular weight
    waxes can be obtained.
3. Direct hydrogenation by primary liquefaction. In some
    cases (such as SRC-2) process conditions have been
    changed from those used initially in order to produce
    lighter products. In essentially all of the primary
    liquefaction processes, some of the light products are
    distillate fuels. Raw oils as primary products may also
    be hydrogenated, reformed, cracked, etc., to produce
    more valuable distillable product liquids.

    With the prospects of the known gasification technology
becoming better, byproducts from the Lurgi, Koppers Totzek,
Winkler, and Texaco processes also become candidates for producing light products from coal. For example, seven byproduct
streams are claimed to come from a typical Lurgi plant. These
include rectisol, naphtha, tar, tar oils, crude phenol, ammonia,

and sulfur [15]. The properties of the light products include
the following:

Tar oil. Specific gravity less than water; contains
water, tar acids, and some nitrogen compounds.

Rectisol naphtha. Specfic gravity less than water,
contains olefins, paraffins and aromatic compounds,
but with significantly more alkylated compounds than
light oil.

Some light products can also be obtained by processing some of
the other crude streams.

In summary, light oils produced from coal must be considered as secondary reaction products. This explains why, as temperature increases, the composition of light oils (and other
products) becomes less a function of the properties of the feed
coal. As the temperature or the reaction time increases, the
yields of light oils and gases increase. This occurs because the
conditions for the cracking reactions that produce these products
become more favorable.

The production of light liquids by hydrogenation processes
is inherently more expensive than for hydrogen. Outside of South
Africa, using Fischer-Tropsch synthesis, no other methods were
operated commercially before 1978. New applications of catalysts,
such as in the Mobil process, or other innovations could make the
production of light liquids such as gaoline feasible but this is
unlikely without the simultaneous production of other valuable
products.

8.4 CHEMICALS FROM COAL

For many years chemicals that have been used for the manufacture of such diverse materials as nylon, styrene, fertilizers,
activated carbon, drugs, and medicine, as well as many others
have been made from coal. These products will expand in the
future as petroleum and natural gas sources become strained to
supply petrochemical feedstocks. The ways in which coal may be
converted to chemicals include: carbonization, hydrogenation,
oxidation, solvent or gas extraction, hydrolysis, halogenation,
gasification (followed by catalytic synthesis), and amination.
In some cases such processing does not produce chamicals in the
sense that the products are relatively pure and can be marketed
as even industrial grade chemicals. The discussion here excludes
processing to chemicals where the products are composed of solutions and mixtures with no clear predominant product. These have

already been covered earlier in this chapter or in the previous
chapter on heavy liquids.

Several authors have reviewed the production of chemicals
from coal and the methods used to produce such products [14-16].
The methods that can be used to produce chemicals include the
following.

Gasification to synthesis gas, followed by catalytic
synthesis (Fischer-Tropsch).

Oxidation.

Carbonization (pyrolysis).

Plasma pyrolysis.

Hydrogenation.

Some of the most important products, and the methods of production from coal are listed.

Aromatic hydrocarbons: BTX, durene, pseudo-cumene, and
naphthalene.

Olefinic hydrocarbons: Ethylene, propylene, butylene, buta-
diene, acetylene.

Other products: Tar acids, hydrogen, sulfur and sulfur
products, hydrogen, sulfur, ammonia (and salts) phthalic
acids, alcohols, acetic acids, and phenolic compounds.

A complete description of the processes to produce all of
the possible chemical products is beyond the scope of this discussion. However, some of the most important processes are discussed.

Table 8.2 gives a partial list of chemicals that can be produced from coal. As indicated in Chapter 6, many primary products
of coal reactions are very large molecules. Secondary products
are often smaller, but become less related to the feed coal as
secondary reaction conditions become more severe (higher temperatures or longer reaction times). Secondary reactions become synthesis, upgrading or concentration reactions of the primary products with hydrogen, oxygen, or other reactants. The basic steps
to most of the products that could be designated as "chemicals"
from coal have already been given. Some of the specific applications of these methods follow.

The importance of chemicals from coal has been recognized
for some time. Particular attention has been paid to the long
range importance of coal as chemical feedstock by such gatherings
as the First International Conference of Chemists (Toronto,

TABLE 8.2   Chemical Species—Candidates for Products from Coal

Inorganic Chemicals	
sulfuric acid	H_2SO_4
sulfur	S
ammonia	NH_3
Organic Chemicals	
benzene	C_6H_6
toluene	$C_6H_5CH_3$
xylenes	$C_6H_5(CH_3)_2$
ethylene	C_2H_4
ethane	C_2H_6
ethylene oxide	$(CH_2)_2O$
propylene	C_3H_8
propylene oxide	CH_3CH_2CHO
alcohols (methanol, ethanol, propanols, etc.)	ROH
butane, butylene, butadiene	C_4H_{10}, C_4H_8, C_4H_6
acetic acid, acetone	CH_3COOH, CH_3CH_2CHO
styrene	$[C_6H_5'CH_2CH-]_n$
phenols	C_6H_5OH and derivatives
formaldehyde	CH_2O
carbon tetrachloride	CCl_4
acrylonitrile	RCN
cumene	
cyclohexane	C_6H_{12}
naphtalene	$C_{10}H_8$
anthracene	$C_{14}H_{10}$
urea	$(NH_2)_2CO$
vinyl acetate	$CH_3CH_2CO_2CH$
phthalic anhydride and derivatives	$C_6H_4C_2O_3$

Canada, July 1978) which was devoted to the sources of raw materials for organic chemicals.[*] While few shortages or crises are expected before 1990, the long lead time necessary for commercial scale development of alternate sources for petrochemicals have made advance planning highly desirable.

Sources of chemicals to take the place of petroleum as this resource becomes more scarce and expensive include tar sands, wood and fast growing plants, organic wastes (primarily agricultural), and oil shale. However, on a world scale, only coal appears to be a long range solution. Coal is far ahead of most alternatives in terms of reserves, availability, commercial acceptance, and technology development. The supply of organic "petrochemical-like" hydrocarbons and related materials will be linked to fuel supplies for the foreseeable future. This also puts coal in a favorable position as a logical source.

8.4.1 Gasification

Synthesis gas ($CO + H_2$) is produced by one of the methods outlined in Chapters 4 and 5. This synthesis gas can be processed in any one (or more) of the following ways.

Reaction over iron catalysts to produce:

    paraffins
    aromatic hydrocarbons
    olefins
    oxygenated compounds (isosynthesis)

Reaction utilizing the water gas shift reaction and nitrogen to produce ammonia [18].

Synthesis over nickel catalysts (methanation--Chapter 5) to produce methane.

Reaction over catalysts to produce methanol (or other alcohols or glycols) [14].

Some of the products listed can be further processed to give other products too, such as methanol, ethylene, •acetic acid, and aldehydes.

---

[*]M. Gellender, "Coal Will Underpin Chemical Industry," Canadian Chem. Proc. Vol. 62, No. 8, pp. 21-22, 1978.

8.4.2 Oxidation

Organic acid compounds (mostly aromatic) can be produced from coal by catalytic oxidation using oxygen or air or by reaction with strongly oxidizing acids. The only practical method of those mentioned is oxidation with air. The products depend considerably on the rank of feed coal since these oxidation reactions are mild enough to give primary products that may be only fragments of the coal structure. Products include benzoic acid, phtholic acid, mellitic acid, isophthalic acid, and tarephthalic acid. Several groups have shown interest in these products although commercial production has not been accomplished [19, 20].

8.4.3 Carbonization (Pyrolysis)

The first part of this chapter was devoted to this process. Essentially all of the commercial carbonization today is done primarily to produce metallurgical coke. The final temperatures of the high temperature carbonization processes used in the range 900 to 1200°C. Because of these severe conditions the products are essentially the same regardless of the feed coal. The most quantities of chemicals obtained are BTX (and other aromatic hydrocarbons), phenols, cresols, and ammonia. These come mostly from the light oil fraction of the carbonization products. In addition high boiling phenols, phenanthrene (from creasote), pyridine, 3-picoline, quimoline, isoquimoline, and naphthalene are also obtained.

8.4.4 Plasma Pyrolysis

By making coal one electrode of an electric reduction process a plasma may be created that reaches higher tempteratures than conventional coal processing reactions [21, 22]. In such an environment acetylene and other small molecules are produced, though this process has not been commercially successful. Acetylene may also be made from coal or coke by reaction with calcium carbide or by a submerged-flame process from coal derived liquids (or from crude oil). Although the use of acetylene has diminished for some chemical because of the advantages of cheaper ethylene and propylene, it is still extensively used in the production of halogen derivatives such as acetaldehyde and vinyl chloride.

8.4.5 Hydrogenation

Generally the oxidation of coal is a process that utilizes the chemical energy of carbon and hydrogen to produce heat. In the production of chemicals by oxidation the thermal energy is lost or inadequately utilized. Hydrogenation, on the other hand, requires a more expensive reactant, but also produces more valuable products. If coal is hydrogenated to produce liquids and gases as fuels, the hydrogenation conditions can also be utilized to produce chemicals such as phenols, tar bases, and ammonium sulfate. Some chemicals are necessarily produced and must be recovered due to environmental constraints; these include sulfur and ammonia. Because of the reducing conditions present, sulfur and nitrogen are recovered as H_2S and NH_3 in the gases from the hydrogenation step. In the case of sulfur, recovery cannot be complete because from 20 to 40 percent of the sulfur is left in the solid residue (coke or char).

Table 8.3 lists typical recovery data for chemicals produced from coal by liquid-phase hydrogenation. The sulfur is recovered by scrubbing gases containing H_2S with alkali solution and burning in a Claus unit. Ammonia is neutralized by passing through sulfuric acid and recovered as ammonium sulfate.

Additional chemicals could be recovered from the hydrogenation of coal if economic incentives were sufficient. Examples include high molecular weight polynuclear aromatics (pyrene, benzopyrene,, coronene, etc.), oxygenated compounds, and high carbon solids (that may be used in reduction processes).

8.5 SUMMARY

Light liquids and chemicals are some of the most valuable materials that can be produced from coal. The economic viability of processing to these products depends not only on the relative value compared to other coal-derived products, but also on the cost of producing these materials from other raw materials, especially petroleum. The chemicals mentioned in this chapter can generally bring a higher price than either light or heavy oils from coal. In a commercial operation the recovery of such chemicals could make economic an otherwise impractical processing scheme.

Predictions of demand for petrochemicals, many of which can be manufactured from coal, show dramatic increases for the future. One study predicts that by 1985 the amount of chemicals produced from coal will be over 8 times the 1975 output (approximately

TABLE 8.3 Typical Yields of Chemical Products from Liquid-
Phase Hydrogenation of Coal

Chemical Product	Weight Percentage of Total Product
Tar acids	
Phenol	1.9
O-Cresol	0.2
m- and p-Cresol	2.4
Xylenols	1.6
Aromatic compounds	
Benzene	8.2
Toluene	13.9
Xylenes	15.4
Ethylbenzene	2.8
Naphthalene	3.7
Other	6.8
Total Chemicals from Liquids	59.9[a]
Other Chemicals	
$(NH_4)_2 SO_4$	11.25
Sulfur	0.72

[a]The remaining 43.1 percent by weight of the liquid and condensi-
ble product was made up of LP gases and gasolines.

1.2 million tons [23]. The realization of such a prediction
depends on many factors. Among these are the price of crude oil
and natural gas, the availability of these resources, and advances
in the processing of coal in commercial scale operations. The
availability of large quantities of certain light liquids and
chemicals from coal depends on the amount of coal processed to
liquids and gases. The specific types of products also depends
on the advances made in knowledge of the chemical structure of
coal and in understanding the ways it can be made to react.

REFERENCES

1 M. B. Sherwin and M. E. Brank, "Chemicals from Coal and Shale," An R & D Analysis for the National Science Foundation, Chem Systems, Inc., NTIS #PB 243,393/AS 442 pp., 1976.

2 J. R. Martin, "Clean Fuels," from Coal Cymposium II, Institute of Gas Technology, Chicago, Ill., June 1975.

3 Minerals Year Book, U. S. Department of Interior, 1959, 1960, et.

4 Benzole Producers, Ltd., Half-Yearly Report of the Technical Department, 1958.

5 H. H. Lowry (Ed.), "Light Oil and Other Products of Coal Carbonization," in Chemistry of Coal Utilization, Supplementary Volume, NAS-NRC Committee on Chemistry of Coal, J. Wiley & Sons, New York, Chapter 15, pp. 629-674, 1963.

6 D. Spencer, Industrial Chemist, Vol. 34, pp. 287-293, 1958.

7 K. H. Y. French, The National Benzole Association, Research Paper 1, 1953.

8 G. Claxton, Benzoles: Production and Uses, National Benzole and Allied Products Association, London, 1961.

9 A. L. Roberts, J. H. Towler, and B. H. Holland, Gas Council of Great Britain, Research Communication GC 31, 1956.

10 J. J. Morgan and R. P. Soule, Chem. Met. Eng., Vol. 26, pp. 1025-1030, 1922.

11 W. Fuche and A. G. Sandoff, Ind. Eng. Chem., Vol. 34, pp. 567-571, 1942.

12 N. A. Gruzdeva, Zhur Priklad. Khim., Vol. 25, pp. 980-993 and 1045-1056, 1952.

13 W. Peters, Schellentgasing von Steinkohlen, Habilitationschrift, T. H., Aachen, 1963.

14 G. A. Mills, "Alternate Fuels from Coal," Chem. Tech., pp. 418-423, July 1977.

15 R. Serrurier, "Prospects for Marketing Coal Gasification By-products," Hydrocarbon Process., pp. 253-257, Sept. 1976.

16 I. Wender, "Catalytic Synthesis of Chemicals from Coal," Catalysis Rev. - Sci. Eng., Vol. 14, No. 1, pp. 97-129, 1976.

17   W. A. O. Herrmann, "Oils and Basic Organic Chemicals from
     Coal by Hydrogenation" (A Literature Review), Canadian Dept.
     of Mines and Resources, Information Circular, IC 229, 46 pp.
     1969.

18   Frank Brown, "Make Ammonia From Coal," Hydrocarbon Process.,
     Vol. 56, No. 11, pp. 361-366, Nov. 1977.

19   R. Smith, R. C. Tomarelli, and H. C. Howard, J. Amer. Chem.
     Soc. Vol. 61, p. 2398, 1939.

20   W. L. Archer, R. W. Montgomery, K. B. Bozer, and J. G. Louch,
     Ind. Eng. Chem., Vol. 52, p. 849, 1960.

21   L. L. Anderson, G. R. Hill, E. H. McDonald, and M. J.
     McIntosh, "Flash Heating and Plasma Pyrolysis of Coal,"
     Chem. Eng. Prog., Symp. Series, Vol. 64, No. 85, pp. 81-88,
     1968.

22   R. Gannon, and V. Krukon, "Coal Process Development," Final
     Report, prepared for the Office of Coal Research, U.S. Dept.
     of Interior, April 1972.

23   K. V. Rao and I. Skeist, Oil Gas J, Vol. 74, No. 2, pp. 90-
     93, 1976.

CHAPTER 9

ECONOMIC AND ENVIRONMENTAL COSTS ASSOCIATED WITH COAL CONVERSION

9.1 INTRODUCTION

For a variety of reasons economic and environmental issues have been consolidated into one chapter, for they help define the limits to employing coal conversion technology. They demonstrate that while coal conversion is beginning to be economically competitive for some industries now, it is not an inexpensive energy solution. These issues also identify both the internal and external costs of coal conversion—costs that must be considered when evaluating nonpetroleum energy options for industry.

Economic issues associated with coal conversion include the ability to obtain coal and water, and the capital and operating costs of conversion facilities. Environmental issues include the effluents discharged from such systems that must be treated. Necessarily both the economic and environmental issues, when viewed as constraints, are heavily influenced by governmental decisions.

9.2 ECONOMIC ISSUES ASSOCIATED WITH EMPLOYING COAL CONVERSION

Since the National Energy Plan [1] and other Federal policies are mandating a conversion to coal, and since sufficient boiler fabrication capacity to handle an immediate massive conversion to direct firing of coal does not exist at the present time [2], the extent to which coal conversion to gaseous or liquid fuels is economic must be determined. Questions of obtaining the raw materials, raising the capital, and affording to produce the gas or purchase the liquid are far from moot. These issues determine the cost of the gas itself. And the costs presented here must be compared not only with current average and more importantly marginal fuel prices, but also with common forecasts that by 1985 energy will cost an average of $\sim\$4+/10^6$ Btu in 1976 dollars.

9.2.1 Raw Materials Availability

Two raw materials, coal and water, are critical to the deployment of coal conversion systems. The extent to which they are available or unavailable sets physical-economic limits on coal conversion systems. Coal, presently, is basically demand limited and will remain so for more than another decade [3]. This could change to a supply limitation, however, if orders for coal suddenly appeared at unprecedented rates. If such orders occurred, for the next 10 to 15 years coal could be supply limited. The coal producing industry must get clear, consistent economic signals from both the marketplace and the government in order to expand beyond presently programmed rates. The lead times required to plan and construct new coal mines are from 5 to 8 years [4]—lead times that require consistent signals. Since such signals have not been forthcoming, it is essential to use modified present growth rate forecasts.

For the past five years coal production grew by 2.24 percent annually—from 595.5 million tons in 1972 to 665 million tons in 1976 [5]. The industry produced $\sim$640 million tons in 1977, a level heavily influenced by the protracted coal strike.) If the trend of increasing coal mining continues in unchanged form, coal production will be 812 million tons in 1985 and 906 million tons in 1990. Present forecasts anticipate a slightly higher growth rate—an annual growth of 4.2 percent. This will produce some 965 million tons in 1985 and 1.2 billion tons of coal in 1990 [6, 7]. Supporting these growth rate forecasts are planned expansions now on the drawing boards or under construction in the western coal fiels. Table 9.1 is an enumeration of these expansion programs as compiled by the U.S. Bureau of Mines. Present expansion plans in the west are designed to add 472 million tons of annual production in the coal industry [8]. It is recognized that not all of these mines will end up being built, as many are dedicated to specific projects. The 965 million ton estimate assumes that about 65 percent of these new mines or expansions will be brought on line, and that eastern coal production will not decrease. New eastern mines will be required to replace worked out deposits, and expanded production capacity from the Appalachian and Illinois coal fields will be limited.

It should be noted that the U.S. Bureau of Mines study shows 125 million tons of planned additional capacity from western coal

TABLE 9.1 Planned Future Coal Mines in the Western States[a]

State	New Mines		Expansions of Existing Mines		Total	
	Number	Annual Production Increase (in 10^6 tons)	Number	Annual Production Increase (in 10^6 tons)	Number	Annual Production Increase (in 10^6 tons)
Arizona	–	–	2	5.0	2	5.0
Arkansas	–	–	1	0.1	1	0.1
Colorado	37	38.8	8	6.8	45	45.6
Kansas	1	0.25	–	–	1	0.25
Iowa	–	–	1	0.1	1	0.1
Montana	6	19.0	3	29.2	9	48.2
New Mexico	3	34.0	5	43.7	8	77.7
North Dakota	5	34.0	4	8.6	9	42.6
Oklahoma	7	0.25	–	–	7	0.25
Texas	5	38.0	1	8.0	6	46.0
Utah	25	54.3	5	10.2	30	64.5
Washington	2	2.0	–	–	2	2.0
Wyoming	23	118.4	10	21.4	33	139.8
Total	114[b]	339.0	40	133.1	154	472.1

[a] Source: [8].
[b] Includes 25 mines in Colorado, Montana, North Dakota, Oklahoma, Utah, and Wyoming with unknown or unavailable capacities.

138

projects is dedicated to coal conversion projects [8]. The magnitude of that commitment indicates that coal can be made available for gasification and liquefaction plants. The Keystone Report for 1976 shows some 24 million tons of expansion capacity dedicated to coal conversion, but even this represents 0.5 quads of energy [3]. Market mechanisms exist for obtaining such dedications, ensuring that coal can be obtained by industries installing conversion facilities at prices estimated to be $\sim$\$0.40/10^6 Btu for western coal and $\sim$\$0.80-1.00/10^6 Btu for eastern coal. Again these prices are in 1976 dollars [9].

Of more immediate significance is the unused capacity in existing coal mines. Suppliers of low Btu gasifiers claim that their units can be fabricated and delivered to using industries within a year after orders for such units [10]. For purchasers of such units, unused capacity in existing mines must be the answer. According to the coal panel of the Committee on Nuclear and Alternative Energy Systems, such capacity could supply over 100 million tons (2 quads) of coal within the existing price structure and an additional 100 million tons of coal with some price increases [7].

While coal can be made available with reasonable certainty, the water question is less clear. Traditionally water is treated as an environmental issue, however its physical availability is of more than passing economic interest. Water can determine what regions are particularly suited for, or constrained from developing coal conversion plants. As has been shown previously, water is both consumed and produced in various stages of coal conversion. Consuming activities of gasification include synthesis gas production, raw gas scrubbing, shift conversion, purification, and all cooling activities. Water is produced as an effluent from scrubbing and cooling, shift conversion, and methanation. The shift conversion, final purification, and methanation phases apply only to SNG production while scrubbing and cooling are related to all gasification systems. In liquefaction water is utilized as a hydrogen source and is also employed for cooling purposes.

The specific water requirements vary by conversion process and individual plant design. The principal variables are the extent to which water is recycled, and the extent to which water is used for cooling purposes. The U.S. Geological Survey shows minimum water demands for SNG production to be 72 gal/10^6 Btu produced using water for 90 percent of cooling. The requirement is 37 gal/10^6 Btu produced using air for 90 percent of cooling requirements [11]. The U.S. Geological Survey estimates of water demand for coal liquefaction suggest a minimum of 31 gal/10^6 Btu produced as fuel. These water requirements contrast with a rate of 146 gal/10^6 Btu produced as electricity from fossil fuel based generating stations [11].

Detailed process water consumption rates have been calculated for specific gasification and liquefaction systems. These are presented in Table 9.2. More general estimates are available for cooling water requirements assuming 75 percent air cooling and 25 percent water cooling. These requirements are presented in Table 9.3. From these data one can observe that to produce 250 × 10^9 Btu in the form of low Btu gas requires nearly 300 × 10^6 gal water,* producing the same amount of energy in the form of medium Btu gas requires some 650 × 10^6 gal, and producing that amount of energy as SNG requires between 560 and 800 × 10^6 gal. It should be noted that the El Paso SNG plant was designed to consume 8.8 × 10^6 gal/day while producing 288 × 10^9 Btu as SNG [12].

To produce 0.1 × 10^{15} Btu from coal at a single plant, then would require 1.6 to 3.2 × 10^9 gal water/yr. This is about equal

TABLE 9.2 Process Water Consumption by Conversion Process (in gal/10^6 Btu output) [12]

Process	Gross Water Consumed	Water Produced	Net Water Consumed
Low Btu Gasification*	6.2	4.2	2.0
Medium Btu Gasification			
Koppers-Totzek	4.9	3.3	1.7
Winkler	13.0	8.3	4.7
Substitute Natural Gas Production			
Lurgi	19.7	16.3	3.4
Synthane	17.4	13.1	4.3
HyGas (steam-iron)	20.0	7.2	12.8
CO_2 Acceptor	12.6	6.3	6.3
Liquid Boiler Fuel Production (Synthoil Process)	3.9	2.5	1.4

*Assumes using the sensible heat in the gas.

TABLE 9.3 Cooling Water Requirements for Coal Conversion[a] [12] (in gal/10^6 Btu output)

Process	Cooling Water Required	
	Make-up	Blowdown
Medium Btu Gasification[b]	20.0	3.8
Substitute Natural Gas Production	16.0	3.0
Coal Liquefaction (for boiler fuels production)	9.4	1.8

[a]Assumes 25 percent water cooling, 75 percent air cooling in all cases.

[b]Assumes cooling and cleaning of gas before combustion. Water requirements for low Btu gas are assumed to be correspondingly less.

to 6,000 to 12,000 acre ft/yr of water. Because of the enormity of such water requirements, it is essential to determine which geographical locations offer sufficient supplies of water for large scale projects, assuming that the needs of small scale plants (∿1 to 100 × 10^9 Btu/day) basically producing low Btu gas can be met in most regions of the country. When determining the locations best suited to large conversion systems, it is also essential to match water and coal availability.

The matching of water and coal availability has been performed by the U.S. Bureau of Mines [13]. Figure 9.1 is their map showing the areas of highest potential in terms of raw materials supply and fuel demand. It illustrates the limiting effect of raw materials availability. For the smaller, single industry low Btu gas plants, the constraints are less than severe since raw materials requirements are lower by at least an order of magnitude.

9.2.2 The Costs of Coal Based Synthetic Fuels

Both the absolute cost of various synthetic fuel options and the relative cost of synthetic fuels vis-a-vis alternatives such as imported oil and gas or electricity are of critical importance in assessing this energy option. Of particular significance is

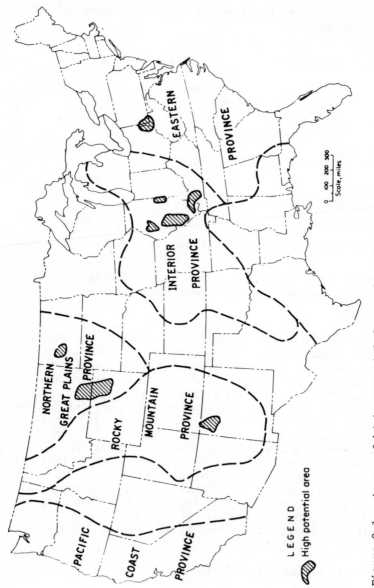

Figure 9.1. Areas of high potential for coal conversion. This map, modified from the U.S. Bureau of Mines presentation, shows where coal and water resources can both be found. The areas of high potential are particularly appropriate for large SNG or coal liquefaction plants.

the fact that all of these can be treated, by the using industry, as marginal costs—the cost of obtaining the next increment of energy supply. Thus they define the real alternatives and trade-offs.

It should be noted that, for the most part, synthetic fuels from coal are economic substitutes for petroleum and natural gas. When the price of oil, for example, rises, demand for oil decreases albeit slightly. The price elasticity, depending upon petroleum product, is in the vicinity of 0.1 to 0.3. When the price of oil increases, the quantity of low Btu gas demanded also increases. Thus synthetic fuels fit the classic definition of substitutes. However synthetic fuels are more than substitutes. They may be viewed to an extent as energy insurance against a shortfall of oil or natural gas. Some portion of their cost may be charged off as protection against a plant shutdown. Thus, if the price of steam generated by burning natural gas or producer gas is equal, there may be advantages in switching assuming all other factors are held constant.

Because synthetic fuels are substitutes for other alternatives, their costs of production can be viewed as marginal costs of new energy supply. Their costs can then be compared to other alternatives such as imported fuels or purchased electricity.

Like all emerging energy alternatives, synthetic fuel systems are capital intensive. The cost of the end product is determined in large part by the magnitude of the initial investment, and the dollar costs associated with paying for and recovering the investment. This, in turn, implies that the total cost per 10^6 Btu produced is heavily influenced by the size, complexity of the process train, and load factor of the synthetic fuels plant. As has been mentioned previously, the cost of these synthetic fuels must be competitive with alternatives such as the direct burning of coal. For any given company, determining the competitiveness of coal conversion involves an individual plant evaluation based upon capital availability and cost, applicable environmental regulations, and other factors such as the cost of raw materials. All of these cost factors are influenced by governmental decisions which help induce uncertainties into the general equations. Because of such uncertainties, only broad estimates of costs are presented here.

Detman has posited a highly useful formula for calculating the constant cost of synthetic fuels [14]. It is based on discounted cash flow (DCF) analysis, and presented here assuming a 12 percent discount (interest) rate.

$$\frac{N + 0.247I + 0.1337S + 0.2305W}{F} = P$$

where I = Total plant investment, initial charge of catalysts
           and chemicals, etc. (millions of dollars)
       S = Startup costs (millions of dollars)
       W = Working capital (millions of dollars)
       N = Total net operating costs in first year (millions of
           dollars)
       F = Total fuel production (in 10^{12} Btu/yr)

The formula assumes a 16 year sum-of-the-digits depreciation
schedule and a 48 percent Federal tax rate. Different conditions
can alter the assumptions, yet such alternatives can be accommo-
dated within the structure of the formula. For example, a manufacturer owning and operating a producer gasifier for its own
benefit might adjust the discount rate in the equation to be
equal to the overall plant discount rate. Similarly the capital
recovery factor may be altered depending on projected plant life
and depreciation practices.

Scrutiny of the Detman formula, or similar analyses based
upon Return-on-Investment (ROI) calculations, shows the intricate
interface between economics and technology. The single most
important term is F, the amount of fuel produced. Fuels from
systems which have unusually low load factors or poor thermal
efficiencies will be inherently expensive. The cost of the fuel
is inversely proportional to F, and in a linear fashion. The
operating cost is of second most significance conceptually, however the fact that nearly 25 percent of the initial capital invested must be paid for annually shows the significance of the
massive capital requirements associated with these plants.

The application of the Detman formula provides reasonable
estimates of the present dollar average cost of various synthetic
fuel options over the life of a given project. Thus the costs of
various synthetic fuels can be estimated for a given moment in
time, for comparative purposes, recognizing certain constraints
on the comparability of such estimates. Those constraints
include 1) different optimum scales of operation, affecting the
level of capital intensivity and resulting unit costs, 2) different ownership patterns associated with the production of various
fuels, and 3) different stages of technology development ranging
from commercial availability for low Btu gas producers to pilot
plants for liquefaction systems. Recognizing these difficulties,
Table 9.4 presents the comparative constant dollar fuel costs
for the varying options. The constant dollar approach has been
used for illustration here as the best approximation of a uniform
comparison of fuels at today's dollar values.

Actual costs will, of course, vary from the estimates in
Table 9.4 due to variations in time specific location, and status
of technology. For example, the table presents an estimated cost

TABLE 9.4 Alternative Constant Costs of Selected Synthetic Fuel Options (in $/10^6 Btu)[a,b]

Cost Component (in millions of dollars)	Fuel Alternative				
	Low Btu Gas	Medium Btu Gas	SNG	Heavy Liquids[c]	Light Liquids
Capital Investment	200	350	1100	800	1250
Start-up Costs	20	20	40	60	40
Working Capital	20	20	40	60	40
Net Operating Costs/yr	70	70	175	195	155
Total Annual Fuel Production (10^{12} Btu)	50	40	95	130	80
Fuel Cost ($/$10^6$ Btu)	2.30	4.10	4.85	3.20	5.95

[a] Assumes 12 percent discount rate; all values in 1976 dollars.

[b] Sources: [9, 14].

[c] Assumes SRC process.

145

of $2.30/10^6$ Btu for producer gas while Ashworth, Vyas, and Bonamer estimate that low Btu gas would cost $2.74/10^6$ Btu. The comparative medium Btu gas figures are $4.10 and $4.66, respectively. While the first figures are average estimates, the Ashworth numbers present the cost if coal conversion were employed in the iron ore industry of Minnesota [15]. Current quotes by Dravo Engineers, The McDowell-Wellman Co., and The Riley-Stoker Co. are at $3.00+/10^6$ Btu in current prices, depending upon the process, location, and coal to be gasified [10].

These data show that low Btu gas, for several companies, is now an advantageous fuel alternative. Thus sales of such units are increasing. In order to solve financial problems of potential customers, Dravo Engineers now offers a package where they will erect and operate gasifiers. The customer purchases the gas rather than making the capital investment.

This competitive position is shown further in Figure 9.2. Low Btu gas is close to the price of imported oil and below the

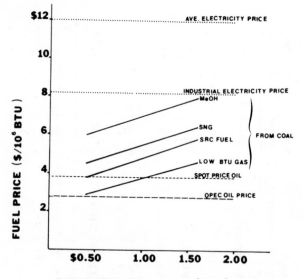

Figure 9.2. The costs of synthetic fuels and some alternatives, 1979. This figure shows that low Btu gas and, to a lesser extent SRC fuels, are becoming economic in areas of low coal costs and appropriate industrial conditions. To the extent that electricity is viewed as the alternative to natural gas, SNG is economic. These lines are used to depict the marginal costs of selected alternatives as industries seek new supplies.

cost of both imported natural gas and electricity. SRC fuels
enjoys a similar position if the cost of coal is held below
$0.75/10^6 Btu. All synthetic fuels enjoy an economic advantage
over the electricity rates picked. (It should be noted that
those are neither the lowest nor the highest power rates available. In Seattle, Washington, electricity costs $4.75/10^6 Btu; in
Chicago it costs $15.96/10^6 Btu; in New York City it costs
$25.60/10^6 Btu; and in Detroit it costs $15.50/10^6 Btu. Such
rates assume purchase of 10,000 kWh/mo. at maximum demand of
40 kW [16].) If one treats the synthetic fuel options as bands
about the lines drawn, one sees the particular applicability of
industrial gas and solvent extraction liquids production.

9.3 THE ENVIRONMENTAL COSTS OF SYNTHETIC FUELS

As has been observed, one reason for utilizing the relatively high cost synthetic fuels is providing for environmental
protection. Conversion is substituted for stack gas cleanup.
The extent of this substitution can be shown with the following
example: combustion of medium Btu gas results in 40 ppm NO_2 and
48 ppm CO released (dry volumetric basis). Similarly, combustion
of natural gas produces 60 and 70 ppm, respectively [17]. Direct
combustion of coal causes additional emissions. Conversion is
not without its effluent control issues. Thus it is important to
consider the production of effluents from selected technologies,
for conversion processes, as well as combustion systems, must
meet environmental standards. Effluents causing both air and
water pollution, plus solid wastes, are considered here as a
result.

9.3.1 Synthetic Fuel Plant Emissions

Coal gasification and liquefaction plants typically generate
significant quantities of waste that must be controlled. The
emissions which must be controlled by air pollution technology
include particulates, sulfur oxides, nitrogen oxides, hydrocarbons, carbon monoxide, and aldehydes. Table 9.5 summarizes the
generation of such effluents by the various processes available.
From Table 9.5 one can observe that the production of 0.1 ×
10^{15} Btu causes the release, hence the need to control, some 400
tons of particulates, 2000 to 8000 tons of sulfur oxides, 3500 to
6000 tons of nitrogen oxides, 80 to 130 tons of hydrocarbons, 250
to 450 tons of carbon monoxide, and 1 to 2 tons of aldehydes if
the process employed is low Btu gasification. If the Fischer-
Tropsch process is employed, up to 1600 tons of particulates,
26,000 tons of sulfur oxides, and correspondingly higher levels

TABLE 9.5 Estimated Controlled Air Emissions From Coal Conversion Facilities[a]
(in lbs/10^6 Btu of fuel produced)

Process/Coal	Products					
	Particu-lates	SO_x	NO_x	Hydro-carbons	CO	Aldehydes
Low Btu Gas (eastern coal)	0.008	0.115	0.084	0.002	0.006	3.1×10^{-5}
Low Btu Gas (western coal)	0.009	0.046	0.135	0.003	0.010	5.0×10^{-5}
High Btu Gas (eastern coal)	0.011	0.149	0.108	0.002	0.008	4.0×10^{-5}
High Btu Gas (western coal)	0.036	0.059	0.173	0.004	0.013	6.5×10^{-5}
Solvent Refined Coal (eastern coal)	0.008	0.044	0.066	0.001	0.006	6.0×10^{-6}
Solvent Refined Coal (western coal)	0.014	0.014	0.075	0.001	0.007	1.0×10^{-5}
Light Liquids[b] (eastern coal)	0.029	0.322	0.270	0.006	0.020	1.0×10^{-4}
Light Liquids[b] (western coal)	0.030	0.127	0.415	0.009	0.031	1.5×10^{-4}

[a]Source: [18].
[b]Assumes the Fischer-Tropsch Process.

of other air pollutants are produced and must be controlled [18].
Air emissions are a function of both process configuration and
process efficiency.

Effluents that can cause water pollution include dissolved
solids, suspended solids, and organic matter (e.g., phenols).
Thermal discharges can also be considered, however these vary
with the extent to which water is employed for cooling purposes
and whether a once-through water cooling mode is employed or not.
Table 9.6 presents representative values for water pollution
problems, exclusive of thermal discharges, for various conversion processes now available or emerging. Again the system efficiency and process configuration determine the extent to which
systems emit water-borne pollutants. It is clear that the low
and medium Btu gasifiers enjoy a significant advantage in this
area of concern.

Coal conversion generates solid wastes, principally ash,
that must be disposed of. This is not unlike handling the solid
wastes from coal combustion. Table 9.7 enumerates representative
amounts of solid wastes generated by various processes. The

TABLE 9.6 Water Pollution Effluents Discharged by Representative
Coal Conversion Processes[a] (in tons/10^{12} Btu produced as fuel)

Process	Effluent		
	Total Dissolved Solids	Total Suspended Solids	Total Organics (e.g., phenols)
Low Btu Gasification[b]	–[d]	–	–
Medium Btu Gasification[b]	–	–	–
High Btu Gasification[c]	43.1	0.9	0.426
Solvent Refined Coal	52.3	0.017	0.003
Light Liquids	63.6	0.008	0.002

[a]Source: [19].

[b]Atmospheric low Btu gasification; Koppers-Totzek medium Btu
gasification.

[c]Lurgi process.   [d]Indicates no appreciable amount.

TABLE 9.7.   Generation of Solid Wastes by Coal Conversion Process[a] (in tons generated per 10^{12} Btu produced as fuel)

Process	Tons of Solid Waste Generated
Low Btu Gasification	7060
Medium Btu Gasification	8430
High Btu Gasification	5270
Solvent Refined Coal	3460
Light Liquids Production	3260

[a]Source: [19].

generation of solid wastes depends, in large part, on the specific coal being processed. Low ash bituminous coals result in less solid waste generation than the higher ash subbituminous coals and lignites. Secondarily, process efficiency determines the amount of solid waste generated.

9.3.2 Controlling Synthetic Fuel Plant Effluents

Employing conversion rather than direct combustion, as has been shown, does not eliminate environmental problems. Rather, it can make them somewhat easier to deal with. In conversion, for example, sulfur tends to combine with hydrogen as H_2S. This contrasts with the formation of SO_2 as a product of combustion. Hydrogen sulfide is more easily removed from the gas stream than sulfur dioxide. Further the volume of fuel gas produced by any process is far less than the volume of stack gas resulting from combustion. Again pollution control is facilitated. The extent to which water pollution control can be accomplished, albeit at a price, can be demonstrated by the Sasol II complex. This Fischer-Tropsch plant now being built has been designed for zero discharge of waste water.

Certain states have formulated air pollution standards for coal conversion facilities, providing benchmarks for industries to evaluate when considering coal conversion. Table 9.8 presents the New Mexico air pollution standards developed in response to the high degree of coal conversion contemplated for that state. For water pollution regulations, Rubin and McMichael suggest that

TABLE 9.8 Air Pollution Standards for Coal Conversion Facilities
Located in New Mexico[a]

Pollutant	Standard	
	Gas Fired Power Plant Associated with Coal Gasification Plant	Gasification Plant
SO_2	0.16 lbs/10^6 Btu	N/A
Particulates	0.03 lbs/10^6 Btu	0.03 gr/scf
Hydrocarbons	N/A	N/A
NO_x	0.20 lb/10^6 Btu	N/A
Sulfur (total)	N/A	0.008 lb/10^6 Btu
Reduced Sulfur Compounds		100 ppm
H_2S	N/A	10 ppm
HCN	N/A	10 ppm
HCl	N/A	5 ppm
Ammonia	N/A	25 ppm

[a]Source: [20].

petroleum refinery or byproduct coking plant standards might be
applied [20]. For solid wastes, current regulations with respect
to coal combustion are likely to prevail.

The standards suggested above can be met for coal conversion
facilities, at a price. It has been estimated that meeting such
standards would require up to 6 percent of the energy produced by
the conversion facility [21]. Mills has shown that meeting environmental standards raises the capital cost of a low Btu gas facility by 20 percent and raises the capital cost of an SNG plant
by 16 percent [22]. Such costs have been included in the estimates presented in Section 9.2.

9.4 CONCLUSION

That the manufacturing community must turn away from natural gas, and ultimately, oil is well understood. Such a reorientation of energy supply is expensive. There are numerous alternatives available to firms moving away from oil and gas. Among these alternatives are the coal conversion options. Low and medium Btu gas systems can be deployed immediately, as can systems for the indirect liquefaction of coal. Other systems which produce liquid boiler fuels will be emerging within the ensuing decade. All of these systems can aid in preserving existing investments (e.g., gas or oil fired boilers), and in meeting environmental regulations with less difficulty than combustion systems.

The use of coal conversion, by any firm, is predicated upon matching up coal and water availability. Such provisions for resource availability can be made. Utilizing coal conversion systems implies accepting increased capital investment in energy systems by the manufacturer or, alternatively, a conversion system supplier; particularly if low or medium Btu gas systems are employed. It also implies accepting higher cost fuels, although the use of low Btu gas provides the least expensive substitute among the synthetic fuels. In effect the capital commitment is traded for the incremental fuel cost advantage when a firm opts for low Btu gas as opposed to SNG or liquids from coal.

The use of coal conversion also involves environmental protection expenditures. Again the capital investment in gasification or liquefaction is traded off against the relative case, if not expense, with which environmental protection can be accomplished.

Today we are in a society seeking every possible alternative to oil and natural gas. It is recognized that for at least the next 50 years, if not longer, coal offers the best general alternative as a substitute. For some firms that implies direct combustion. For many companies, however, that coal alternative can best be employed if the coal is converted into gaseous or liquid fuels. And, for a certain population of manufacturers, the coal alternative can only be employed if synthetic fuels are available.

REFERENCES

1 Executive Office of the President, The National Energy Plan, Washington, DC: U.S. Government Printing Office, April 1977.

2 Office of Technology Assessment, U.S. Congress, Analysis of the Proposed National Energy Plan, Washington, DC: U.S. Government Printing Office, Aug. 1977.

3   Martha Krebs-Leidecker, Synthetic Fuels From Coal, in Project
    Interdependence: U.S. and World Energy Outlook Through 1990,
    Washington, DC: U.S. Government Printing Office (for the
    Congressional Research Service, Library of Congress), Nov.
    1977.

4   Walter Dupree, Hermann Enzer, Stanley Miller, and David
    Hillier, Energy Perspectives 2, Washington, DC: U.S. Govern-
    ment Printing Office, June, 1976.

5   Commodity Data Summaries 1977, Washington, DC: Bureau of
    Mines, U.S. Department of the Interior, 1977.

6   J. Bhutani, et al., An Analysis of Constraints on Increased
    Coal Production, McLean, VA: Mitre Corp., Jan. 1975.

7   Report of the Coal Sub-Panel to the Supply-Delivery Panel of
    the Committee on Nuclear and Alternative Energy Systems,
    National Academy of Sciences, 1977.

8   John Corsetino, Projects to Expand Fuel Sources in Western
    States, Washington, DC: Bureau of Mines, U.S. Dept. of the
    Interior (I.C. 8719), May 1976.

9   John Phinney, "Basis for Projections," Background paper for
    the Supply-Delivery Panel of the Committee on Nuclear and
    Alternative Energy Systems, National Academy of Sciences,
    1977.

10 Paula Ruth, "Coal Gasifiers Experiencing Renaissance," in
    Energy Users News, Vol. 3, No. 20, May 15, 1978.

11 George H. Davis and Leonard A. Wood, Water Demands for
    Expanding Energy Development, Washington, DC: U.S. Geologi-
    cal Survey (Circular 703), 1974.

12 Ronald F. Probstein, et al., "Water Needs for Fuel-to-Fuel
    Conversion Processes," presented at the 67th Annual Meeting,
    American Institute of Chemical Engineers, Washington, DC,
    1974.

13 A. E. Lindquist, Siting Potential for Coal Gasification
    Plants in the United States, Washington, DC: Bureau of Mines.
    U.S. Department of the Interior (I.C. 8735), 1977.

14 "Coal Gasification Commercial Concepts: Gas Cost Guidelines,"
    Project 4568-NW, U.S. Energy Research and Development Admin-
    istration, Washington, DC, Jan. 30, 1976.

15 R. A. Ashworth, K. C. Vyas, and D. G. Bonamer, Study of Low
    and Intermediate Gas from Coal for Iron Ore Pelletizing,
    Cleveland, OH: Arthur G. McKee & Co. (for U.S. Bureau of
    Mines), March 1977.

16 "Commercial Electricity Prices by Cities," Energy Users News, Vol. 3, No. 20, May 15, 1978.

17 Ad Hoc Panel on Liquefaction of Coal, Assessment of Technology for the Liquefaction of Coal, Washington, DC: National Research Council, National Academy of Sciences.

18 Synthetic Fuels Commercialization Program, Vol. IV, Draft Environmental Impact Statement, Washington, DC: U.S. Energy Research and Development Administration, Dec. 1975.

19 Energy Alternatives: A Comparative Analysis, Norman, OK: The Science and Public Policy Program, University of Oklahoma, May 1975.

20 Edward S. Rubin and Francis C. McMichael, "Impact of Regulations on Coal Conversion Plants," Environmental Science and Technology, Vol. 9, No. 2, Feb. 1975.

21 R. M. Jimeson, "Utilizing Solvent Refined Coal in Power Plants," Chem. Engr. Progr. 62(10) pp. 53-60, 1966.

22 G. Alex Mills, "Synthetic Fuels from Coal: Can Research Make Them Competitive?" presented before the Washington Coal Club, Washington, DC, March 16, 1977.

INDEX

155